Happy to Work Here

Understanding and Improving the Culture at Work

Steve McMenamin
Tom DeMarco
Peter Hruschka
Tim Lister
James Robertson
Suzanne Robertson

```
Copyright © 2021 Steve McMenamin, Tom
DeMarco, Peter Hruschka, Timothy R.
Lister, James Robertson, and Suzanne
Robertson

ISBN: 978-0-9892820-6-2

Cover by James Robertson. Design and
interior illustration by James Robert-
son. "Brit and Italian" graphic on page
4 by Tom DeMarco. Continuous line il-
lustrations by ngupakarti. Author pho-
tos by the authors.

Version 1.7c: April, 2021
```

All rights reserved. Neither this publication nor any part thereof may be reproduced or transmitted in any form without the prior written permission of the copyright holders. The copyright holders' consent does not extend to copying for general distribution, for creating new works, or for sale. Specific permission must be obtained for those purposes.

Please direct all inquiries to cultureproject@systemsguild.com

Visual Table of Contents:

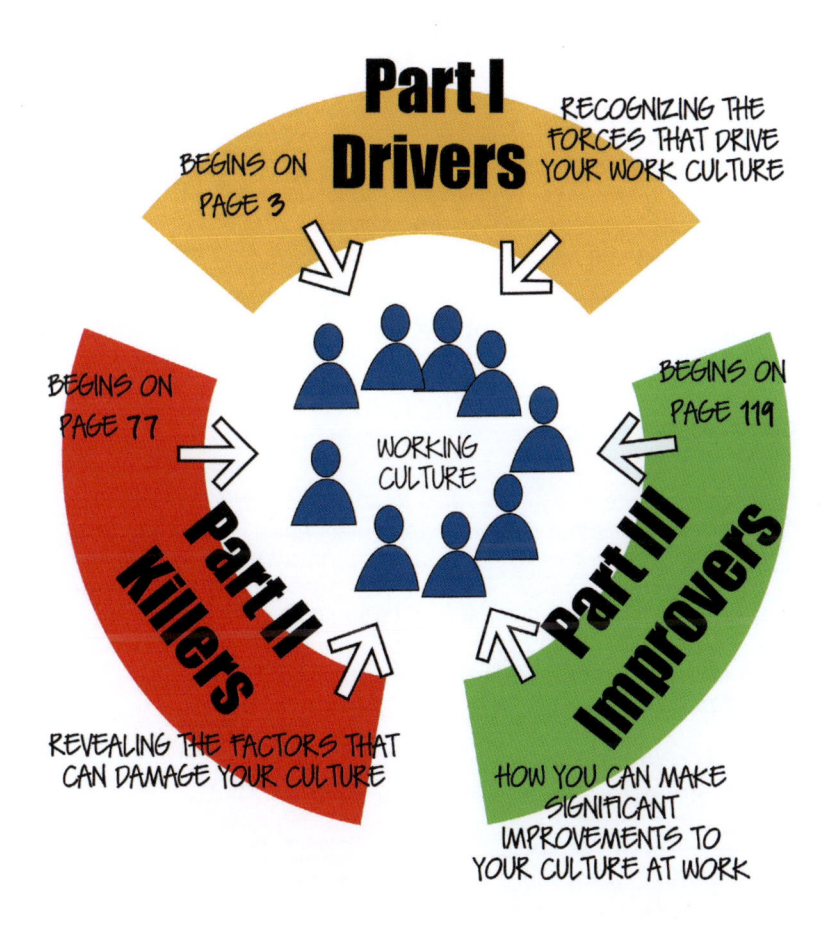

Part I — Drivers
BEGINS ON PAGE 3
RECOGNIZING THE FORCES THAT DRIVE YOUR WORK CULTURE

WORKING CULTURE

Part II — Killers
BEGINS ON PAGE 77
REVEALING THE FACTORS THAT CAN DAMAGE YOUR CULTURE

Part III — Improvers
BEGINS ON PAGE 119
HOW YOU CAN MAKE SIGNIFICANT IMPROVEMENTS TO YOUR CULTURE AT WORK

Early Praise for *Happy to Work Here:*

To me, the key word is "Happy." If you're not happy where you are, who you're with, what you're doing, or where you're going - then you'll have to alter your course. Understanding the culture that we're immersed in or responsible for, is the first step in creating a safe and sustainable place where we can self-actualize and create a life of meaning.

And that's the point - achieving our best selves, whether as individuals or as a collective is often elusive. But not so elusive if you read this book. Highly recommended!"

–Michael Mah, Managing Partner, QSM Associates

The book seeks to be practical. It is written by practitioners for practitioners. With language that practitioners can understand. So, if you're looking for evidenced academic theories and insights into the complexities of organization culture, don't buy this book. If you are looking for a simple read that provokes you to ask straightforward questions about your places of work in order to make projects go better, then buy it. It provides an insightful, experience-laden introduction to a complex but important subject.

–Dr. Neil Maiden, Professor of Digital Creativity, Cass Business School, City University of London

While there are many frameworks for understanding culture at an overview level, the means for actually changing culture is less well understood. This book provides extensive insights into the killer behaviors leaders should watch out for and the drivers where cultural change efforts should be focused. It is an essential read for anyone wishing to understand a prevailing culture, who refuses to accept that this is just the way things are and wants to know how to make a difference.

–Dr. Debra Paul, Managing Director, AssistKD, London

The Working Culture You Want

You can *sense* culture at work. When you first come into contact with an organization, a company, a department, a division, or a team, you get an immediate impression of its working culture long before you can articulate why you feel the way you do about it. Sometimes you sense the crackle of energy of purposeful people enjoying their work and doing it well. Other times you just know that the culture is so toxic that if it weren't for the paycheck, nobody would show up.

Culture matters. It is fairly apparent that high performing organizations have favorable cultural profiles. The people enjoy working for the organization and are pleased with, and proud of, the quality of their work. They feel that the organization values them, and in return do all they can to produce valuable work products.

> *One of my neighbors runs a small consulting firm that does a culture screen of companies to guide potential investors. His clients are convinced that a strong, confident culture is a leading indicator of organizational success, while financial results are only a trailing indicator.*
>
> *— Tom DeMarco*

Poorly performing organizations most often have cultures that are part of the problem. Culture influences performance, and performance influences culture. The connections between culture and performance are complex, reciprocal, and anything but obvious, and you need to care about both. The more you care about performance, the more you need to care about culture.

Anyone who is responsible for a complex team endeavor or continuous process is aware that the methods in use, the technology, and the work ethic don't provide a complete explanation for why people and teams perform as they do. There is something else at work; that something else is its culture.

You often hear allusions to it: "The culture of this place just didn't allow us to pull off that initiative," or "It was the culture that lost us some of our best people." But when asked to explain just what the culture is, or even what a phrase like "our organization's culture" really means, most people come up blank.

> *Seeking to achieve good culture and avoid the bad requires a deeper look at the factors that determine working culture. That's what this book is about.*

Today's fluid work environments make working culture harder to assess, but even more important. We used to bump into workmates when we went for coffee in the office kitchen. We would chat about work and about personal things. These personal chats over coffee lubricated our communication and made us feel closer to our workmates.

However, an employee in temporary workspace that will be hers only for the duration of the current project, or the one working at home, or the guy in the office whose teammates live in another country, still have the same cultural needs. Wherever they are they have to feel valued, they need to have the same amount of socializing and interpersonal interaction as they had in the kitchen. They need the same understanding of the goals and the direction. They need the same sense of safety and belonging as anyone working in co-located space. The need for good working culture does not go away if you move out of joint space, but it may take more effort to get it right.

This book presents a distillation of many working cultures, especially including some admirably healthy ones. We hope you'll recognize some aspects of your own culture in these discussions and gain useful clues about how to improve it.

A look ahead

The six chapters of Part I present the drivers, the major factors influencing workplace culture. Part II consists of short essays describing some common culture killers – toxic factors that damage a culture, sometimes beyond repair. Part III is our prescription for the daunting yet doable task of culture improvement.

Part I
The Drivers of Working Culture

Culture is a set of shared beliefs and resultant observable behaviors.

Before getting deeply into that idea, consider for a moment its *story*, the story that tells of how we first came to believe in that idea. As we were beginning this book project, we came across a startling observation by the researcher Edward T. Hall from work he did back in the 1960s. What Hall observed was that people have different notions of personal space, different ideas for what constitutes "too close" in a face-to-face interaction, as well as "not close enough" to communicate comfortably. Hall went on to assert and prove that these proper distancing beliefs were common within one culture, but different for different cultures. So, for example, Italians share a belief that a comfortable conversing distance is pretty close to the face of their listeners, while British people share a belief that more space is required between them. The result is that an Italian and Brit conversing together at a cocktail party might find themselves moving slowly in a characteristic way with the Italian feeling compelled to move forward to achieve acceptable distance, while his

British friend retreats. Probably neither is aware of the slow dance they're doing:

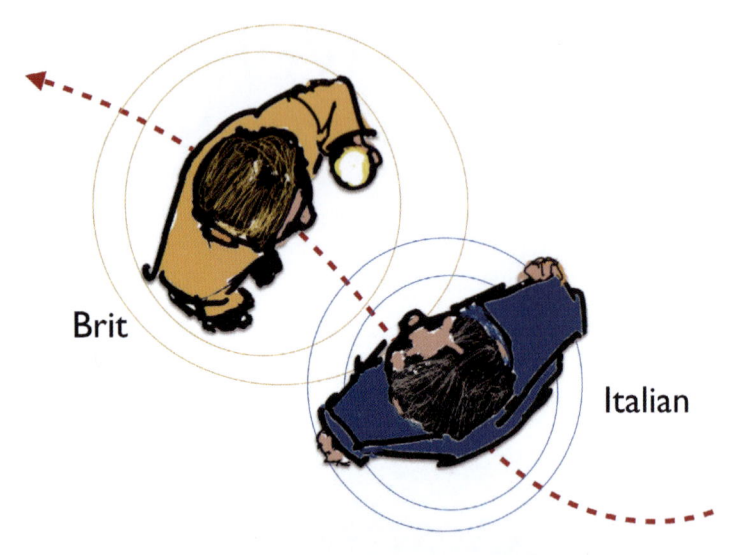

Brit **Italian**

Could it be, we wondered, that this is not just one random characteristic of a culture, but a prototype of all the things that make up that culture? Would it be possible to *define* a culture in terms of such shared beliefs? That was the genesis of this book.

Shared beliefs are fundamental – they determine behavior. For example, if you believe that the world will end next Tuesday, you will behave quite differently than you would if you were confident your existence and that of the planet would continue. People in Denmark believe that most other people can be trusted which causes Danes to behave in a respectful manner towards each other. Compare that to the behavior of people who believe that their fellow humans are vile, untrustworthy, deceitful villains. People who believe that they might be fired soon behave differently from those who believe they are highly valued by the organization.

In our work experience and observations of hundreds of organizations around the world, we have, naturally enough, encountered a variety of cultures. The shared beliefs that define a culture are seldom immediately apparent; they need to be *inferred* from analysis of observable behaviors.

For example, when people share the belief that people themselves are the prime assets of the organization they tend to behave in a respectful, willing, and collaborative way towards each other. In organizations where navigation – the forward planning and setting of clear directions and objectives – is believed in by all, people tend to behave cooperatively and calmly. They know where they are going, and what they have to do to get there.

There are a number of these combinations of shared beliefs and resultant behaviors. We have called these combinations *cultural drivers*.

The drivers are shown as an aggregate of cause and effect in Figure 1.1.

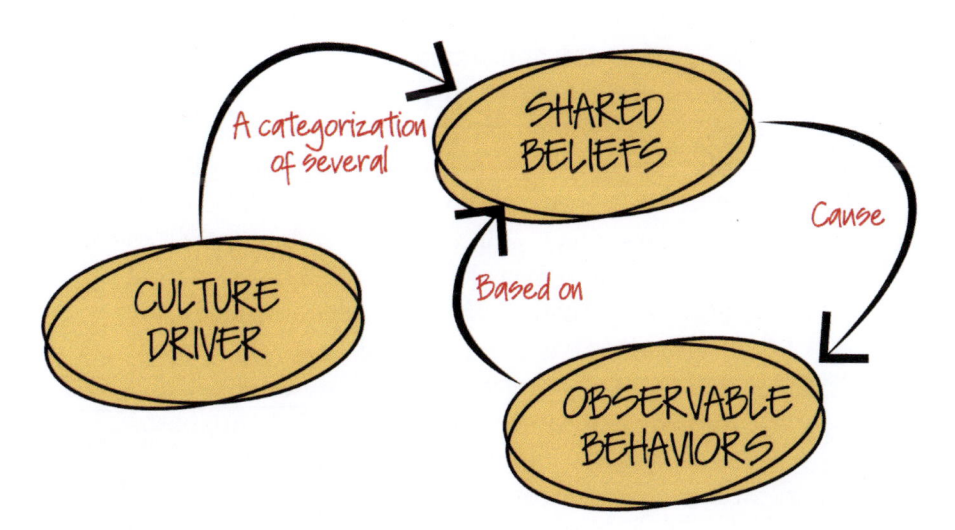

Figure 1.1 Each driver of working culture comes about because of one or more shared beliefs. The existence of these beliefs can be inferred from analysis of observable behaviors.

Six Drivers

We have identified six culture drivers, and while there may be more, these are the ones that if acted upon, will have the most beneficial effect on your culture. They are set out below:

1. ***The perceived value of people and teams****:* How are people seen in the organization? Are they seen as highly valued assets to be selected and invested in to further the organization's true objectives? Or are they viewed more like lightbulbs: commodities to be used until they burn out, and then replaced with cheaper ones? Are they valued enough so that even if they are isolated and working from home, management goes to some lengths to ensure their comfort and help keep them feeling like part of the whole? Are work teams persistent? Do they interact more like basketball teams, or more like golf teams?

2. ***The perceived nature of time:*** How is time perceived by the people doing and managing the work? Do they generally accept that adding time to a task can increase the probability of success? And contrarily, that extending the duration of a task exposes it to increased costs and the risk of disruption from external and internal changes? In this respect, time can be both a friend and an enemy. Do they see that time is a scarce and valuable resource and understand that wasting time is a crime? Are they always aware of the importance and value of time? Are they given sufficient low-level clerical and administrative support so that highly skilled specialists are not wasted doing their own?

3. ***Safety and security:*** Do people feel comfortable that they won't be demeaned or degraded? Are they made to feel safe and part of the team even when remote from their workmates? Is there a willingness to take on risk knowing that some failure is tolerated as

the price of finding better, more innovative solutions? Is there a willingness to trust fellow workers and the organization? Do managers trust their remote workers to get on with their work without constant monitoring and micromanaging? Are managers intent on enabling the growth and advancement of their subordinates without feeling threatened by the teams they manage?

4. ***Navigation by grown-ups:*** Is the business vision embodied by a set of prioritized intermediate and long-term goals that everybody knows and understands? Are the goals realistic and are the schedules for achieving them based on reality? Is the organization moving steadily towards these goals and not being frequently diverted by short-term emergencies? Does the organization prioritize and focus on a few strategically important initiatives? Can it decide what not to do?

5. ***Collective confidence:*** Are workgroups confident that they are collectively capable of overcoming problems they might encounter, and so take on more challenging activities? Do they focus more on wins, or more on avoiding losses? Are they afraid of risk, or do they see risk as a necessary part of innovation and advancement? How do they respond to uncertainty? Do they freeze with indecision, or does their collective confidence allow them to be undaunted by uncertainty? Are they collectively confident even when they are dispersed?

6. ***The perceived value of excellence:*** Is there an evident pride of workmanship? Do people ever use the word "beautiful" to describe their product or service? Is the organization driven to turn out exceptional products and services? Given monetary and market constraints, does it do its utmost to achieve its full potential: to be as good as it can possibly be?

Of course, these six drivers cannot account for all of the rich variation among working cultures, but they account for most of it. By focusing on the drivers and their contribution to good cultural mores, you are more likely to understand the culture around you. Understanding the six drivers, detailed in the rest of Part I, should help you to move towards the working culture you want.

Chapter 1
The Perceived Value of People and Teams

If "happy to work here" is a common attitude in your workplace, it's a strong indication that the prevailing culture places a high value on people, their skills, talents and characters. This value is real, it is far above insincere declarations and motivational posters – it is palpable. You can sense it in the way that the organization is careful to hire the right people, and in how employees are respected and treated. You can see it in the way that people respond to being valued.

In this, the first of the culture drivers, we look at an organization that genuinely places a high value on people and the teams they work in.

Scenario: Interview day

You have been invited to interview for a new job. If you get the job, you will be working with a group of people that you have held in high regard from afar. You met the person who invited you at an industry conference about a year ago. Over the following months, you have spoken several times and gotten to know each other. About a month ago, he introduced you to a colleague, who is the hiring manager for the position you are seeking today. After a lengthy video interview, this manager invited you to come to meet the team and sit for some additional interviews.

You arrive at the company's offices and are greeted by the manager, followed by coffee and small talk, since this is the first time you have met in person. Eventually, he outlines the day he has planned for you. It will be a series of interviews, each lasting thirty to sixty minutes, and it will probably take all day. You will start by meeting some of the team members who will be your peers if you get the job. Afterward, you will have a longer discussion with the manager, who would be your direct supervisor if you decide to join the team. After that, you are told, it is hard to say – several other team members might ask if they can meet with you.

What follows feels like a marathon that lasts until it is getting dark outside. Your first three interviews are with members of the team that you are seeking to join, several of them communicating remotely. All three seem to you to have similar attitudes to their work and the organization. They also appear to have been together in the team for some time. Each of them has a copy of your resumé, and clearly has read it. But beyond asking a few "ice breaker" questions, they don't seem to pay much attention to it. They mainly pose specific work scenarios to you and ask you how you would approach them. They appear to want to understand how you think, and how you communicate. One of them even presents you with a "brain teaser" puzzle and asks you to describe how you would solve it.

Your fourth interview is with your prospective manager. He queries you about how you see your past experience enabling you to do the job you are applying for. Some of his questions however are more general and appear to be aimed more to your potential future in the company than to your initial involvement. He follows up on a couple of the points that were covered in the first three interviews. Next, you interview with one of the hiring manager's peer managers. After that, the team's Human Resources representative meets with you briefly to address any questions you might have about the company, its policies, benefits, and hiring processes.

Your seventh interview of the day is bit of a surprise. You are greeted by the Senior Manager, to whom both the hiring manager and five other managers at his level report. She asks some general, and somewhat off-the-wall questions that have little to do, in your view, with the details of the work you are applying to do. She also gives you the chance to ask any questions you have about the company, its strategies and plans, and about the role you're seeking. She is friendly, but very matter of fact. This last interview lasts barely thirty minutes. She then returns you to the office of the hiring manager. He leaves you alone for a few minutes, and then returns to tell you that everyone was very impressed with you, and that they would like you to join the team, provided of course that you and the company can arrive at a mutually satisfactory agreement on salary, benefits, etc. You leave their offices exhausted, elated, and a bit bewildered. You almost wonder whether you are joining a team or a cult.

You might be even more bewildered if you knew what had been going on behind the scenes throughout the day. Each person who interviewed you was given a copy of your resume ahead of time, along with a list of everyone on the interview loop (yes, even including the Senior Manager). After each interview, the person who had met with you would *immediately* post his/her thoughts in a collaborative document shared with everyone else on the loop. He or she would recount the high and low points of your discussion with them, along with their assessment of your strengths and weaknesses, and, if any, their recommendations for topics later interviewers might address if they have time. Finally, each interviewer would boil down his/her reaction to you in a single sentence: "Hire." or "No hire." In making this judgment, they were not telling the manager what to do. They were simply declaring what their verdict would be if the entire hiring decision rested upon them.

Your second interviewer was meeting with you while the first interviewer was posting his thoughts in the collaborative document, so she didn't see it until later. But the third interviewer had read the result of the first interview before he met with you. The hiring manager was following all the interview results throughout the day. While the HR rep was meeting with you, the manager was assessing how you had done. Since you were doing extremely well, he contacted the Senior Manager and asked her to meet with you also. (She had already reserved the time.) If you had blown it, she wouldn't have been called in.

While you were relaxing after the final interview, the manager was making a final canvas of all the interviewers and running his planned decision ("Hire") past his boss to make sure that her talk with you had raised no red flags.

What did your interview day tell you about this company?

A community hiring process sends the very clear message that you've been considered and evaluated not just for your individual skills and qualities but also for potential fit with the group. It gives you every reason to believe that the people who took part in the interviews came to view you as someone they'd like to work with, and also someone who is likely to complement and strengthen the team. For this organization it takes a community to be sure it has the right person for the long haul.

The approach was clearly orchestrated. You can't be sure about all that went on behind the curtain, but you know as a certainty that you did not meet with seven random people who were told this morning that they each would interview a prospect.

The people you met were serious, relaxed, and confident. These are the hallmarks of people who are and feel trusted.

You are living proof that it is not so easy to gain admission to this company. At the most obvious level, the employees of this company have demonstrated that they believe the hiring process is very important. Look at the level of planning and effort they are willing to invest in a single candidate. Look at the breadth of direct participation in the interviews themselves. In many companies, the interview would have been conducted only by the hiring manager and perhaps someone from HR. In this case you had those two participants, plus three team members plus three levels of management, including a manager from an entirely different part of the organization.

The care and precision the company put into your interview were very like that which would precede a serious investment. The interview day was an investment, and a sign that this is the kind of organization that is inclined to invest in its people.

Taken all together, your interview reveals the cultural characteristics of an organization that clearly:

- strives to hire only the best people it can find
- declines to compromise on the above

- works to preserve harmony and well-being of the team by avoiding poor hiring decisions
- knows it is better *not* to hire someone than to hire the wrong person
- trusts its people to assure the future health and well-being of the enterprise

All of these are indications of a set of shared beliefs. Shared beliefs are of vital interest here as they determine the working culture. Here we see a culture at work that holds the following belief:

SHARED BELIEF *The proficiency of its people is the most critical contributor to the organization's success over time.*

And, since these people can only achieve meaningful results when working together, fitting into the team is as important, or even more important, than fitness for the job:

SHARED BELIEF *The organization values effective and harmonious teams over individual performance.*

Observable behaviors arising from these shared beliefs

These shared beliefs spawn behaviors that you can see throughout the organization. The approach to hiring that you experienced during your interview marathon was one such manifestation. There are others:

There is real investment in the long-term development of people

The organization does not hire people just to do the jobs they interview for – it recruits people for the long term, expecting that they will stay for years, and that their roles will change over time as their preferences and abilities, and the workplace needs, evolve.

Consistent with this expectation, the organization sets aside funds for the ongoing training and development of its people. These are not the first funds to be sacrificed to a budget-cutting jihad; they are viewed as a high priority, like employee compensation. Moreover, training and development investments are not limited to skills that are directly applicable to the employee's current assignment. More fundamental skills and capabilities that will foster career growth are also approved routinely and without discussion or need for defense.

Development of people is not restricted to career needs – human needs are equally important. HR (remembering that the "H" stands for human) can be mobilized to provide guidance on organizing work and family matters – work hours, moving and settling, schooling for the children, and so on. The organization is strongly aware that its people are human and have everyday human needs.

Teams are tight-knit and persistent

Team composition is often the result of natural selection. Team members have loyalty to one another; they respect one another. Each knows that the others have passed the same difficult tests to get into the organization in the first place. They are willing to listen to one another and interact without pretense or posturing. Team members know exactly how good their teammates are and treat them accordingly.

This does not mean that there are no disagreements or arguments; there surely are. But when team members disagree, they take issue with each other's ideas, not with each other as people. In fact, once team members have achieved a solid level of mutual trust, confidence, and respect, it actually enables more candid, and less guarded dialog among them. It also means that team members can choose to stay together for several, perhaps many, projects. Their management naturally welcomes the persistence of these teams.

Peer coaching is part of everyday work

Teammates understand that nobody knows everything (they are no longer teenagers). The team members respect knowledge, and so are always trying to increase their own, and their teammates' knowledge and skills. They are never afraid that a teammate would use superior knowledge against them.

The team takes pride in turning out successful products or services

This common purpose brings teammates together. This respect and loyalty make a successful team behave much like a successful family.

Peer managers exhibit team behavior

Rather than see other managers as competition and a threat, managers who believe in the value of people cooperate and collaborate with their peers. Peer managers provide counsel and guidance to each other, step in to cover absence, work together to develop schedules and procedures, and are content to coach and be coached by other members of this management team.

Management team members know that they all work for the same organization. They celebrate successes regardless of whether the success is from their team, or a peer manager's team. They have a sense of joint ownership of projects, activities, schedules, and work plans.

The organization goes to great lengths to accommodate talent

The organization seeks to accommodate (within reason) a candidate's preferences for facilities and equipment. When people are forced to work from home, the organization is willing to supply funds to make their workspace at home a little better than the kitchen table. In some special cases, it is willing to alter its basic work practices in order to gain essential talent. For example, a newly hired employee who was initially expected to work in the main office, is allowed to stay remote to accommodate his spouse's work needs. Doing this is unusual, but everyone understands why it makes sense to do it. The organization is demonstrating its belief that the employees are a long-term asset and should be kept for many years.

There are limits to these accommodations. The organization is tolerant of the exotic personalities that sometimes accompany game-changing talent. But the one thing not tolerated is an employee whose behavior is harmful to the team. A good deal of the effort invested in the hiring process is designed to avoid toxic hires. When one slips by, it is dealt with swiftly, regardless of the talent lost. The harmony and effectiveness of the team matter more.

> *We had a team of about forty programmers developing a large C++ application. Some of them were outstanding programmers. There was one guy in the team who outperformed many others in terms of the quality of his code, the robustness, the elegance, and the speed he could create it. But one of his habits was to change any piece of code that he considered suboptimal by rewriting it from scratch. Over time his colleagues started to dislike this practice more and more – especially when their code got changed even when it worked. They secretly had to confess that the new solution was better in many respects, but it was often hard to read and understand since the genius did not bother to document the new parts; he considered his code to be self-explanatory. The company finally decided to get rid of that genius despite his skills.*
>
> — *Peter Hruschka*

Recognition and compensation measures are tailored to individual employees. When the time comes for the annual salary increase exercise, the company tries to determine how best to spend its limited funds, so it delivers as much perceived value to the individual employees as possible. Sometimes this involves rethinking some established norms:

> *Our boss at an Australian software service bureau was very pleased with Barry's work. Everything he worked on was*

innovative, on time and on budget, and valued. We all loved working with Barry on the networked accounting system project. The boss wanted to show his appreciation, so he promoted Barry to be the Development Manager. He got a new office, new responsibilities, and a significant rise in salary. There was only one problem: Barry turned out to be a lousy manager.

We complained of course; we wanted our valuable work colleague back, not a poor manager floundering in his new role. The boss was worried that Barry's salary was above normal for a developer, but none of us minded. The outcome was that Barry got his old job back (and kept his new salary). We got our guru back and the replacement development manager was actually paid less than Barry. It was all public and nobody objected to people being paid for their experience rather than their job titles.

— Suzanne Robertson

The connection between shared beliefs and observable behaviors

Any organization that genuinely believes in the value of its people and teams would display some, perhaps all, of the behaviors listed above. Naturally enough, believing in its people and teams makes for a better culture, but it also makes the organization an attractive place to work. This in turn lures lots of talented job candidates, which in turn means that you can keep the barrier to entry high. You don't worry so much about the odd false negative if there is an unending stream of high-quality candidates lined up at your door.

When the organization has an entirely different perception of the people it employs

Sadly, all-too common, are organizations that exemplify the very opposite of this cultural driver. It is worth taking a moment to look at these kinds of organizations, where:

- Prospective job candidates have to survive a gauntlet of HR interviews and are subject to endless forms and tests before they even get to meet with the person to whom they would report. They don't meet their new team members until they show up on the first day of work.
- Job offers are based on a corporate standard offer for each position. The terms of the offer cannot be tailored to emphasize the elements that are important to individual applicants.
- Respect for teams, teamwork, and team chemistry is mostly lip-service.
- Performance reviews are used to rank-order employees. (Look out if your score is on the low end.)
- In theory, there is a specific amount budgeted for ongoing training and personnel development. In reality, these are among the first funds to be cut during each year's budgeting process even though most employees place a very high value on professional development.
- In times of austerity there is a limited salary increase budget. Usually this means that employees who are not on the cusp of being terminated should get a raise that they won't find insulting. So, if the overall salary increase pool is 3%, just about everyone is expected to receive a raise of at least 2.5%. This means that the world-class superstar who became an employee as a result of a recent acquisition cannot receive more than a 3.5%, regardless of the magnitude of his or her contributions. After all, if such people leave, you can always find someone (probably at a lower salary) to replace them.

What are the shared beliefs that drive such behaviors? We can start by saying that *these organizations see employees as the equivalent of furniture*:

1. Employees are, like furniture, a regrettable necessity. You need a certain number of them to get the work done. But you wouldn't want to have more than you absolutely need. And you wouldn't want to pay them any more than you have to.

2. The choice of one job candidate over another is not going to affect the company's ability to succeed in its missions.
3. Employees are interchangeable. As long as they meet the minimum requirements for their position for knowledge, skills and experience one employee is as good as another. If one leaves, they can be quickly replaced.
4. Since employees are merely tactical assets like office supplies, the goal is to minimize the total cost of ownership of the human assets.
    a. When hiring, pay as little as the chosen candidate will accept.
    b. Avoid increasing salaries unless you have no choice.
    c. Adopt plans for your office space that enable you to house as many employees per square foot as local fire and health regulations allow.
    d. Promote competition among your employees. Make sure they know that they need the company much more than the company needs them, and that successes experienced by their peers are a threat to their continued prosperity.

Understanding and improving your working culture

We leave you to look into the beliefs of your own organization. Are people believed to be the crown jewels of the organization or are they merely furniture? It is by influencing and changing these beliefs that you can have an impact on behavior. Understanding the beliefs that prevail must be the first step in improving culture.

Chapter 1 Visual Summary

SHARED BELIEF: PROFICIENCY OF PEOPLE IS THE MOST CRITICAL CONTRIBUTOR TO SUCCESS OVER TIME

OBSERVABLE BEHAVIORS:
- RECOGNITION AND COMPENSATION ARE TAILORED TO INDIVIDUALS.
- CANDID & CONSTRUCTIVE FEEDBACK ON PERFORMANCE.
- THE ORGANIZATION GOES TO GREAT LENGTHS TO ACCOMMODATE DESIRED TALENT.
- REAL INVESTMENT IN THE LONG-TERM DEVELOPMENT OF PEOPLE.

EFFECTIVE AND HARMONIOUS TEAMS ARE VALUED MORE THAN INDIVIDUAL PERFORMANCE
- PEER COACHING IS PART OF EVERYDAY WORK
- TEAMS ARE TIGHT-KNIT AND PERSISTENT.
- THE TEAM TAKES PRIDE IN MAKING SUCCESSFUL PRODUCTS & SERVICES.
- PEER MANAGERS EXHIBIT TEAM BEHAVIOR.

Chapter 2
The Perceived Nature of Time

If the talent and effort of its people fuel an organization, time is the oxygen that mixes with the fuel to propel the organization forward. To succeed, you need to combine the right amount of time with the right mix of people.

This driver looks at the nature of time as a resource, and how the working culture is affected by the way it is perceived.

Time as a resource has two contradictory characteristics. The first is that adding time to any activity is one of the most powerful tools for increasing the probability of its success. Time is your friend. The second is that adding time is also likely to increase the probability of failure, as we shall see later in this chapter. So, time is also your enemy. Therefore, every time-allocation decision is a double-edged sword: adding time means more time to successfully complete, but also increased exposure to risks associated with extended project time.

Why time is your friend

The benefits of allowing sufficient time are evident:
- People given enough time are more likely to produce a complete, thoroughly thought through, high-quality product.
- Having the right amount of time enables the team to search for superior solutions without being pushed to rush to the next task.
- Sufficient time allocation will probably provide the opportunity to incorporate input from a broader range of stakeholders.

Spending time to better understand the requirements, for example, reduces the chance of functional gaps that later lead to unplanned additional work to compensate.

I recall the best manager I worked for: John, a manager at a magazine publishing business, was a master of time. He understood that it was pointless to waste time and knew how to make the best use of it.

John's progress reports were always pessimistic, and in several cases, he altered my numbers, so it appeared that less progress had been made than was actually the case. When I questioned this practice, he told me that the other managers put in breathlessly optimistic progress reports, and seeing them, the bosses at head office thought they had extra time and piled in with more (often unnecessary) requirements. These additions usually caused those projects to run late. 'They never give me extra work like that' said John, 'but I always complete my projects on time.'

— *James Robertson*

Why time is your enemy

You are well aware that the longer a project (or other activity) takes, the more it will cost. Most projects have costs that are overwhelmingly driven by human effort and its supporting costs like facilities, development tools, training, etc.

Then there are other costs that you might incur if you are thinking of extending the time allocated to an activity:

- Increased duration creates exposure to changes to the business environment that further increase costs and duration. These are changes imposed on the business over time by the environment in which it operates. These could be changes to the rules that govern the environment, or strategic changes adopted by an external company that require compensating changes from your organization, or the withdrawal of support for a key infrastructural component that requires you to modify your interfaces. These environmental changes can affect the business model, or the technology infrastructure, or both.

- Environmental changes happen continuously throughout the life of the business process. Extending the duration of a project by one month will incur one month's worth of environmental changes that – at the very least – must be incorporated into the new and improved business process. Doing so adds still more time to the duration of the project.
- Extending project duration also exposes the project to *internal changes*. If a project to improve your business processes is planned to last for 18 months, for example, you must assume that the people affected by the business process will need to continue to refine it while the new business process is being developed. They cannot simply wait for your results. The longer the new business process is in development, the greater the drift of the new business system from the current business process reality. This means you must spend extra to keep the system-in-development synchronized with evolving requirements, and the longer the project will take.
- If there is a sound business case for your project, then you expect benefits to be realized once the new business process is implemented, but not before. If your project takes fifteen months instead of ten, then you will have incurred five months of deferred benefits opportunity costs.
- Extended duration of a project also exposes the organization to costs outside the domain of the project. If the current project is extended, the benefits associated with the next project in line are also delayed. In this way, project duration exposes the organization to portfolio-delay opportunity costs.

Perceptions of time – the good, the bad and the clueless

You will observe differing perceptions of time in different organizations. The good organizations appreciate that time is to be treated carefully, as both a friend and an enemy as we explained above. Then there are those that are obsessed with minimizing project duration to the detriment of other considerations. These organizations are overly fearful of running out of time. Then there are the completely clueless ones which place no emphasis on managing time – work and projects are allowed to meander on endlessly.

These three types are discussed below:

The Good: An organization that has a balanced view of the dual nature of time

Constructive beliefs about the nature of time assure a well-balanced view of both the positive and negative effects of spending time on an activity. An observer of a team working in such an organization would get an immediate sense of its hustle. There are observable behaviors within the team which include:

There are no leisurely meetings

The team hates to spend time in meetings, especially those that are not necessary. Some necessary meetings are long, but most are short and crisp.

The main work during meetings is the allocation of resources to project activities

These could be newly identified activities, or the allocation of additional resources to activities that were not completed in the anticipated time. Either way, most such proposals to add or extend time for an activity receive withering scrutiny, and many are rejected. The team is very miserly with time.

But not always

There are a few types of project activities that don't get decided so quickly. By unstated agreement among the team members, these activities receive a disproportionate share of the team's discussion and deliberation. Which activities these are varies by the type of project. They might include: (1) conceptual ideation of the project's goal; (2) discovery of the business requirements; (3) end-to-end testing of the solution; and (4) preparing the business organization to succeed with the new process when it goes live.

The team practices relentless prioritization of the project tasks

Team members know that all tasks are not equally important. Some activities, for example those listed above, are more valuable to the success of the project than others. The team is surprisingly willing to add time to high-priority activities if progress lags.

Team members value not only how well they execute the project, but also how quickly they do so

This rapid operational tempo seems to say that it is not enough to implement a good solution, it is equally essential to do so with alacrity. The team seeks to deliver not only quality, but also value. This extends to getting products out the door by the (real, not imagined) deadline.

This same ethic extends to (or perhaps originates with) individual performance

Team members gain respect from others when their work exhibits not only quality but also deftness.

Management reinforces these values by providing adequate low-level support

Skilled workers are kept available to do the work they're best able to do, and not spending time on unimportant administrative chores.

> *The new CEO of Bigelow Laboratory for Ocean Sciences told me that from her first day on the job she was aware that her scientists were spending far too much of their time doing their own low-level support work. She decided to spend some of her "honeymoon capital" to hire clerks and lab technicians to free professional employees to concentrate on what their skills and*

> *training made them best at. Just as laying off support personnel tells people that their time is not particularly valuable, paying for more and better support sends the opposite message, that their time is hugely valued.*
>
> — Tom DeMarco

You can probably already see some of the shared beliefs that underlie the observable behaviors listed above. Here are a few:

SHARED BELIEF *All increases in duration expose the project to additional risk of failure. So, it's no surprise that the team believes that it is a crime to waste time.*

Team members are constantly looking for tasks that can be started *now*. Why hold a meeting just because the schedule says to? *It is a crime to waste time.* Why wait until the end of the meeting to start on the action items? *It is a crime to waste time.* Why drop the third performance test cycle altogether? *Because it is a crime to waste time,* and the current results from the second performance test cycle show that the product is already close to release criteria.

However, there is more than the belief in not wasting time.

SHARED BELIEF *Some project activities are more likely to repay the investment of additional time.*

The team knows that time can also be your friend.

Why extend the system integration test cycle for another four weeks? Because the current test results indicate that this is a necessary investment. Why agree to delay the go-live date for two months to allow time for selective remedial training for users? Because the organization readiness assessment indicated that key users lacked the knowledge of the new business process necessary for them to succeed.

In these cases, team members are still very much aware that their decision to add time to the project increases risk, but this is offset by their understanding that some project activities benefit by being given additional time.

This balanced appreciation for both the positive and negative effects of adding time to a project allows team members to accept potentially unpopular decisions like delaying the system go-live date by two months. They have the understanding — and the language — to express the risk trade-off that supports their decision.

> *The team I was working on at a department store was charged to do the preliminary analysis for a new project. There was a two-year time limit, and we were told that the entire project must be completed on time. But the more we looked into the project the more we realized that it was not possible to meet the deadline.*
>
> *Our team leader pointed out that the preliminary work we had done allowed us to identify the functional chunks of the problem. Initially, the business managers were adamant that all the functionality had to be completed within the two years, despite that now we had identified all the functionality, it was looking increasingly unlikely. We changed the approach by asking the business managers to help us prioritize the functional chunks from a business point of view. In the process of doing this, we discovered that the two-year limit had been set because a new store was planned to be opened by that date. We also discovered that some of the chunks were needed to open the store, but others could be progressively introduced later. With the help of the business managers, we were able to allocate the available time to maximum advantage.*
>
> *Instead of time being a constraint hanging over us like Damocles' sword, we could now see time as a currency, something tangible that we and the business could choose how to spend.*
>
> — Suzanne Robertson

The Bad: Organizations that place too much importance on tight deadlines.

Consider what it would be like if the team's culture focused excessively on the risks of adding time to project activities. Such teams can become obsessed with minimizing project duration over all other considerations. This leads to a variety of harmful behaviors:

- Setting unrealistic expectations. Some team leaders believe that it is good practice to set ridiculously aggressive "stretch goals" in a misguided attempt to ensure that people stay fully engaged
- Refusing as a matter of principle to extend time. This happens when only the downside of adding time is considered, denying the potential benefits of allowing extra time. It is also harmful when it's abundantly clear that the time originally allocated was insufficient.
- Compromising other release criteria. When the team hits a time crunch for integration testing, and the only alternative is to delay the implementation date, a team is often tempted to relax pre-established release criteria to allow a less-well-tested product to be released.

These behaviors – and others like them – diminish the culture, and the behavior, of the project team. Poor time-related decisions mean that people are prevented from doing their best work; this usually results in failed, or vastly inferior, project outcomes.

The Clueless: Organizations that appreciate only the benefits of additional time.

From time to time, but fortunately not too often, we have encountered organizations that do not place *any* apparent value on time. Time-oblivious organizations tend to have poor cost accounting practices, and very poor benefits management practices. They literally do not know what their current practices cost them. They have even less of an idea what benefits will accrue as a result of spending the time to implement a new feature or business process. They continue to add time to a project even when the cost of the additional time outstrips the benefits to be gained from it.

There are two red flags that herald the presence of such thinking. The first is that the business cases address costs but not benefits. This happens when (1) the proposed change has no benefits; or (2) the benefits cannot be tracked; or (3) benefits cannot be harvested because they just wind up in someone else's budget.

The second red flag is a team asking for more time to complete a project, and the manager's response is anything like, "Sure, take all the time you need." The manager is clearly in no hurry to deliver a result since it will probably not contribute to margin either by increasing revenue or by reducing costs.

And time can do so much

Judicious use of time contributes to a harmonious culture. Using a balanced appreciation of both the positive and negative effects of spending time on an activity leads to a more constructive culture:

- Extending time on appropriate activities can lead to delivering better, more innovative, and more complete products. People like to be proud of what they produce.
- Not extending time makes sense when it enables you to make early deliveries in a timely manner of product versions which might not yet be fully complete but perform well enough to be immediately valuable to its audience.

Balancing beliefs in both the benefits and risks of time investments is one of the most important contributors towards the working culture you want.

Chapter 2 Visual Summary

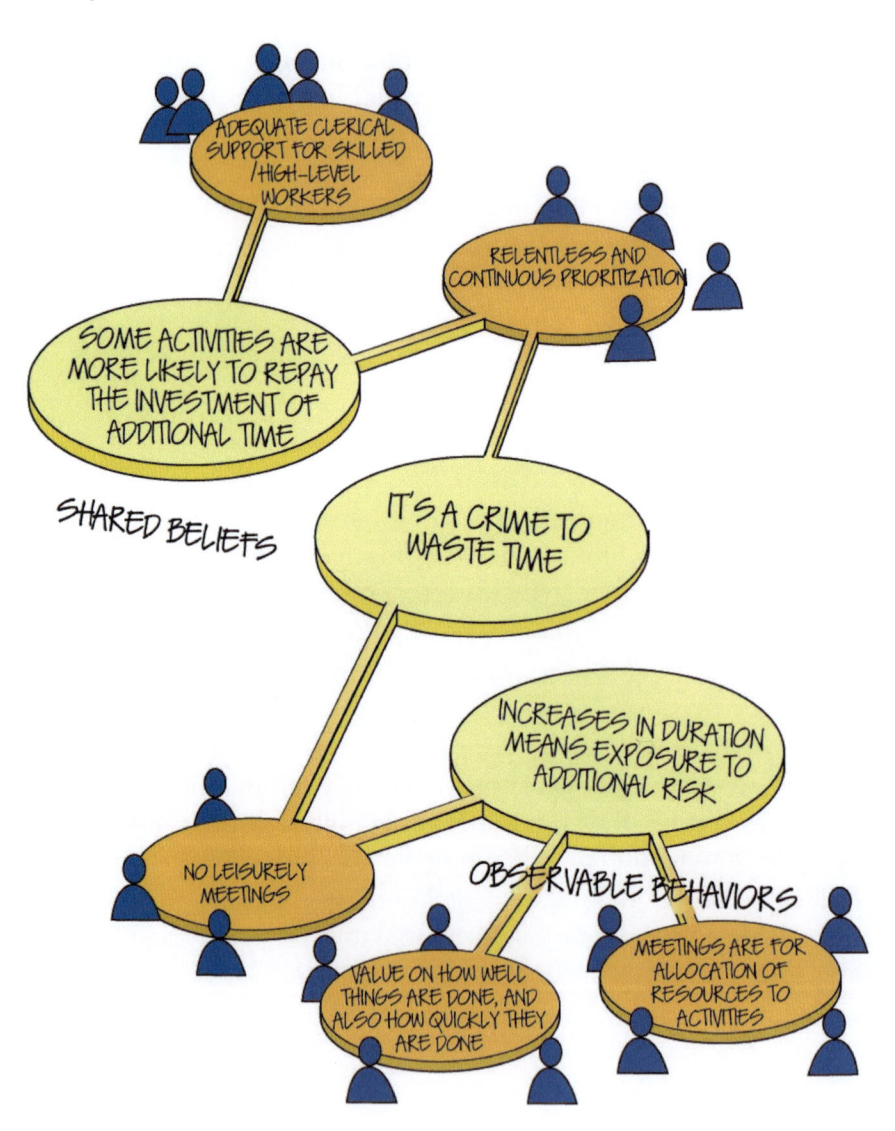

SHARED BELIEFS

- ADEQUATE CLERICAL SUPPORT FOR SKILLED/HIGH-LEVEL WORKERS
- RELENTLESS AND CONTINUOUS PRIORITIZATION
- SOME ACTIVITIES ARE MORE LIKELY TO REPAY THE INVESTMENT OF ADDITIONAL TIME
- IT'S A CRIME TO WASTE TIME
- INCREASES IN DURATION MEANS EXPOSURE TO ADDITIONAL RISK

OBSERVABLE BEHAVIORS

- NO LEISURELY MEETINGS
- VALUE ON HOW WELL THINGS ARE DONE, AND ALSO HOW QUICKLY THEY ARE DONE
- MEETINGS ARE FOR ALLOCATION OF RESOURCES TO ACTIVITIES

Chapter 3
Safety and Security

As a driver of working culture, safety and security play a huge part. Consider their absence: If people are regularly bullied, if management is scornful, full of bluster, and ready to blame everybody, the feel of the organization – its culture – is likely to be dismal. People may have to work in such an organization, but nobody is enjoying the experience. Truly safe organizations share the following belief:

SHARED BELIEF *Nobody here is going to be demeaned or degraded. Ever.*

Safety and security are not the same. You feel safe when you're completely comfortable that you'll be treated with respect by colleagues at all levels. You feel secure if your paycheck is not at risk. There can be no happy culture without safety, but it is possible to find reasonably acceptable situations that aren't entirely secure. Workers in the most wild-eyed you-bet-your-mortgage-on-this-one startups may be at some peace with the uncertainty they have to endure: it's part of the deal. The rest of the deal (fascinating work, chance of riches and renown) might make up for the potential downside. And it helps that everybody is in the same boat: if the boat goes down all hands are equally awash.

While a lack of security is not a fatal flaw for everyone, for most people, most of the time, it is. If you have dependent family, a home and mortgage, or if you work in a locale with few backup job possibilities, or in a declining job market, at least some job security is essential. And safety is always essential.

Bullying

In one of our management courses, we queried attendees if they'd ever observed (either in their present company or another) anyone being bullied in the presence of co-workers. More than half the attendees replied that they had been present in at least one such incident. Our finding is somewhat anecdotal, but data collected more formally by the Minnesota Association of Professional Employees and the Workplace Bullying Institute supports a similar conclusion about U.S. workers at least: "61 percent of Americans are aware of, or suffer from, abusive conduct in the workplace." That amounts to nearly 100 million working people.

Bullying doesn't just happen *down* the hierarchy. It's possible to bully peers, even those higher on the org chart if they're weak enough, and particularly if they're also being bullied by those beside and above them. The effect of being bullied is a sense of estrangement, and when it happens in front of colleagues, humiliation. Although cyber-bullying using the company's chat or email is not always visible to colleagues, it too is distressing and demeaning.

Bullying at work is complex, and there is not much agreement on why it happens. Without blaming the victim, it seems clear that some people are more likely to be bullied than others. For example, sometimes people who are doing outstanding work are bullied. This might be because of jealousy, or people feeling threatened that their own work might be seen as inferior. People with insecurities or neuroses or vulnerabilities are often bullied: the bullies sense this vulnerability, and much as a dog senses fear, home in for the kill. Anyone who differs markedly from any prevailing norm may be subject to bullying. Of course, bullying makes for a toxic culture.

When people feel confident about safety, fellow employees will support the victim, and provide active discouragement to the bully. This means that victims must feel that it is safe to talk about their experience to their colleagues. When it's a superior doing the bullying, it takes courage to stand up for the victim. However, failing to do so assures that the culture suffers a body blow.

Managers in safe organizations act swiftly to stamp out bullying whenever it comes to their attention. Naturally enough, bullying can be no part of a safe work experience.

Fear at work

Fear can be thought of as organization-instituted bullying and can be a dominant force in the organization. People know that their fearful culture makes it an awful place to work, but they also know that, as in North Korea, people disappear if they dare speak up against it.

Fear is the antithesis of trust. Where people are trusted, they are allowed, indeed encouraged, to get on with their work pretty much as they see fit to do it. In a non-trusting, fearful environment, management dictates how and when work is to be done, and weaponizes goal setting as a way to micro-manage the work. These artificial goals are easy to measure, and anybody who does not achieve them (no matter how ridiculous or irrelevant they are) will probably disappear. Everybody knows this, and so the scheme helps create the climate of fear. What's present in safe organizations and absent from culture-of-fear organizations is the following belief:

SHARED BELIEF *Colleagues are (generally) deserving of trust.*

Safety fosters openness. Openness means that people talk to each other and are unafraid of saying what they think (within reason). People naturally enough disagree with each other on work-related matters. One person thinks that opening new shops during a recession is poor policy, another thinks it's a great opportunity to get cheap rental and readily available labor. These views are expressed openly, knowing that these are business matters, and knowing that there will always be several viewpoints but only one can be accepted. Everyone knows that the people who put forward the non-accepted proposals will not be ridiculed or punished.

> *One good practice in agile projects is replacing boring weekly or monthly status reports with daily standup-meetings. Every team member briefly answers three questions: What did you do yesterday? What are you planning for today? And what is hindering you in your work?*
>
> *In one of my projects about half the team members were very hesitant to pick up this new procedure – mainly because of the*

> *third question which involves confessing problems. But it only took four or five days before even the shy ones discovered that their reports are really valuable and valued by the others and — even more important — that answering question 3 is a very non-threatening way to immediately get help from someone in the team. Or at least trigger some action from management to help them quickly overcoming such impediments.*
>
> — *Peter Hruschka*

Openness also means that people can talk frankly with their boss. They are inclined to give better advice when they know the boss is actively paying attention to them. People know they can say it out loud when they perceive that a wrong direction has been taken, or a better solution is available. Openness brings with it more effective communication, vital to the success of any endeavor. Conversely, fear makes communication dangerous and liable to arouse suspicion.

Fear is contagious. People are fearful if they believe themselves likely to be bullied or abused. And fear may also be part of the cause of bullying behavior. In the worst cases, fear works its way up and down the organization chart until it results in a total "culture-of-fear" organization.

Safety, the cardinal rule

There is a cardinal rule about creating and maintaining a positive organizational culture: Bullying and fear must be no part of its makeup. It would be nice to think of that as a prescription: banish bullying and fear from your working environment and you'll have a happy culture. Would that it were so simple.

Culture-of-fear organizations — at least when fully infected — are probably never going to be cured of their affliction. If you're currently working in a culture of fear our best advice to you is to leave. If you realize the need to leave, please do it before you yourself catch the infection.

When people *don't* cut their losses in culture-of-fear organizations, it's often because they are persuaded that things are just the same everywhere, that all organizations are similarly shaped by fear. That simply isn't true. There are fearless organizations all over the world. If yours isn't one of them, it's time to look elsewhere.

The word *fearless* in English has two slightly different meanings: it means literally "without fear," sensing nothing to be afraid of, and it also means "courageous," rising above any fear that one might be expected to feel. To build this second kind of fearlessness into your culture you need to focus on the first kind. People in fearless organizations are ready and eager to take on challenge, something that would be highly unusual in a culture-of-fear organization.

Safety and risk

It's possible to succeed in the short term without taking risks, but not in the long. To succeed in the long run a corporation, a government, or department, or agency needs to progress; it needs to build better solutions to its business problems. However, this kind of progress involves taking some risks.

The kid learning to skateboard knows he is taking a risk. He knows that he is going to fall off his board a bunch of times before he becomes really good at it. But he takes the risk because he wants to make progress – there is no other way. Organizations take risks when they install new systems, or open new branches, or expand into new areas. They know that if they want to progress and grow, they'll need to court risks rather than run away from them.

Risk-taking is, by its very nature, a flirting with failure. An organization that encourages risk-taking but is intolerant of failure is a logical absurdity. And yet it is a common one; we see it everywhere. If even an occasional failure results in punishment of some form, people will become increasingly risk-averse.

To encourage real risk-taking, and real progress, people need to understand that occasional failure is inevitable, and thus expected, that succeeding all the time would be a clear signal that it was just not taking on enough challenge. They need to be encouraged to believe:

SHARED BELIEF *It is safe to take on risk, even to fail sometimes, in the face of serious challenge.*

It is difficult to celebrate a failure, but not impossible. If you're a parent, you know this already: The way you buck up a kid who has tried hard to do something hard and come up short is the very way we need to respond to at least some failure at work. The lesson of celebrating a "good" failure feeds the ability of the organization to grow and change.

Perhaps the best-known example of "good" failure is Thomas Edison. Edison built a business that continued to grow long after his lifetime. The Edison business was based on innovation and inventions. He and his engineers at Menlo Park knew that failure was part of the process and was therefore expected. Edison himself mused tongue-in-cheek that he never failed but had come up with 10,000 things that did not work. Edison's engineers were not afraid of failure; they knew that it was a necessary part of the process of being successful.

> *When Tim and I were involved in an extended consulting engagement at Hewlett Packard we heard of something neither of us has ever encountered elsewhere: an annual prize given for Constructive Disobedience. Each year the winner was someone who had taken the risk of doing some bit of unauthorized work that turned out to provide invaluable assistance to the work that was authorized. The award was given in a jokey ceremony by a senior executive who never failed to state unequivocally that disobedience was not to be encouraged even though the ceremony explicitly encouraged a bit of it.*
>
> — *Tom DeMarco*

When failure is dangerous to the career there is a disinclination to take risks. But sometimes risks must be taken, no matter the disinclination. In that case, the sense that failure is dangerous has the perverse effect of making effective risk mitigation almost impossible. The heart of risk management is mitigation: "We fully expect to win the coming litigation, but here are three things we're doing now to keep options open just in case we lose." Whatever your biggest risk is – loss of an important client or contract, scandal or PR embarrassment, late project, unexpected loss of a key participant – some mitigation in the present can help if the risk does indeed materialize. But if any failure is likely to incur blame, why bother to mitigate? Effort you spend on a just-in-case plan is wasted. Failure leads to punishment, so mitigation makes no one a hero. In fact, in organizations where failure is simply not tolerated, mitigation is seen as pointless, a sign of corrosive defeatism.

Some thoughtful risk-taking is encouraged and admired in safe organizations. You may even be chided (gently) for a long series of successes, a suggestion that taking a few more risks might be a good idea.

When competence is dangerous

If you're a manager, you should hope that some of the people working for you are better at some things than you are. In fact, *everyone* working under you should be better at some things than you are. This is particularly true when the work requires a hybrid mix of skills.

SHARED BELIEF *No one is threatened by excellent performance either below them in the hierarchy or at a peer level beside them.*

If you're threatened by the competence of people who report to you, they are sure to detect it. And then they will begin to understand that competence is dangerous, that displaying too much competence makes them unsafe. They'll take steps to conceal their competence. And how does one conceal competence? That's easy: by acting incompetently. Not good for you. Feeling threatened by competence of subordinates usually results in your subordinates becoming incompetent.

A special case of perceiving competence as a threat is a manager's discomfort with teams. The loyalties that bind teams together may be stronger than the ones that bind team members to the organization and its management. When the people below you have a strong bond to each other, then the work will thrive. You, as their manager, will obviously benefit from the communal bond, but you may have to keep reminding yourself of that.

The way to build an organization where competence is non-threatening is to make a big deal over the manager's role as nurturer. A manager whose subordinates go on to bigger and better things is seen as a success. A manager who cultivates strongly jelled teams is a hero. Part of the success equation for managers at all levels is the extent to which people and teams under them prosper. Conversely, a manager who fails to nurture people's careers so that they rise in the organization is viewed as simply too self-protective.

The mysterious matter of trust

Some organizational cultures are inherently more trusting than others. Specifically, in some the default is to trust while in others the default is to withhold trust until trustworthiness has been definitively proved. Trusting cultures tend to feel healthier and happier than less trusting ones. Micro-management becomes rare, and commitments made by your workmates come to feel as good as the ones you make yourself.

An earlier cultural driver spoke about the organization's perception of people, and how when people are valued, the organization goes to considerable trouble to hire the best people available within the constraints. Having hired those people, it is just common sense to let them make most of their own decisions and judgments. When managers have people working remotely, those people work more diligently when they are given their tasks and trusted to get on with it. They don't need, and certainly don't want constant monitoring. However, it remains the duty of a manager to be concerned with and to make frequent inquiries as to the well being of his/her subordinates. These inquiries are particularly important when team members are working remotely, and the manager is not regularly meeting them during the workday.

Trust means that instead of people working to a rule book, following a rigid procedure, or reading their directions off a screen, they are encouraged to think for themselves. In our experience, people value trust and empowerment over almost anything else. Career growth and training usually comes in second.

Even the most trusting organizations, however, understand that it can be a disaster to place trust in someone not yet ready for it. If a newly hired person is not yet familiar enough with the organization's products and systems to make the right decisions, then complete trust cannot yet be advanced. The objective of a trusting culture is to grant trust *slightly* in advance of demonstrated capability.

There are situations where people can only be given a limited amount of trust. One example is traders working in financial organizations. If a trader "goes rogue," that is, he starts chasing losses by making bigger and riskier bets, it is possible to bring the entire organization down (this has actually happened). These organizations need to have systems that carefully monitor traders to ensure no-one can make potentially cataclysmic deals.

A side effect of advancing trust is the likelihood that you will gain the trust of those you've trusted – think back to how *you* felt about someone who trusted you to achieve what was for you a personal-best stretch goal. The rule of "Give trust to gain trust" is particularly important for managers, since gained trust is an important part of the toolkit they need to manage effectively.

> *The tech leads at a software products company got together to see if they could converge on a specific solution to the problem at hand. I watched as diagrams were scribbled onto whiteboards, debated, erased, and new diagrams appeared. About an hour and a half into it, with no end in sight, the project manager got up, and said, "I can't help here. Keep going, and when you have found the solution, come and get me." He departed. Everyone in the room smiled. The manager had signaled, "This is a tough technical problem, and tough technical people are going to solve it. I trust you."*

— *Tim Lister*

Observed behaviors in a safe/secure culture

In cultures that feel totally safe and reasonably secure, the following are behaviors you're likely to encounter:

Peer coaching
People feel strongly positive about both coaching and being coached. There is a sense that upping the game of anyone in the team is of benefit to all.

Pushback
Dumb decisions meet with objection. What might seem insubordinate in a less safe organization just seems like good hygiene in a safe culture.

Fewer departures
People are more inclined to stay. They enjoy and talk about the healthy culture that supports them.

Investment
Because employees tend to stay, they are viewed as human capital, and the organization is eager to invest further in them. This investment takes the form of training, and/or diversity of job assignment (to round out skills), and/or some risk-taking to advance enough trust for people to have a chance to be challenged.

Dialog about ideas
Criticism takes the shape of a dialog about ideas, never as an attack on the people who advance them.

Trusted workers
People are given their assignments and trusted to get on with it — they are not expected to report back daily. They are trusted that when things go wrong, they are honest enough and feel safe enough to go back to their boss or colleagues and ask for help immediately.

Focus on careers

A job in the organization is not just a job; it is part of a career. Team members are conscious of both their own and their colleagues' career paths. Helping someone along feels good to everyone. And having a good culture means that people think of their work as a career, not just a job. Sometimes the company invests in people's careers by accommodating their specific needs:

> *Annie was one of the best programmers I ever worked with. She was fast, effective, and had an enviable track record for solving difficult problems. Annie was not perfect — she had trouble making it into work before noon, and if she did manage to be there in the morning, she was lethargic and unenthusiastic. The afternoons and evenings were her time. It was an obvious and welcome change when her manager accepted the inevitable and told Annie to start her workday at noon. For the rest of us, we quickly became used to not seeing her in the morning but enjoying her great work in the afternoon and evening.*
>
> — Suzanne Robertson

Safety and success

Safety in an organization almost always means openness and better communication. It means acceptances of ideas; it means understanding and accepting that real progress usually comes with some stumbles along the way. Safety means that people are more interested in making advances than they are in covering their rear ends. Given sufficient and appropriate talent, the result is success in the marketplace.

That sounds like a virtuous cycle, and it is, though beneficial effect on the culture tends to lag behind the success.

Chapter 3 Visual Summary

SHARED BELIEF

- NOBODY IN THE WORKPLACE IS GOING TO BE DEMEANED OR DEGRADED. EVER.
  - CRITICISM IS A DIALOGUE OF IDEAS, NEVER AN ATTACK ON PEOPLE
  - BULLYING HAS NO PART IN THE CULTURE
  - PEOPLE ARE TRUSTED

- NO ONE IS THREATENED BY EXCELLENT PERFORMANCE EITHER BELOW THEM IN THE HIERARCHY OR AT A PEER LEVEL BESIDE THEM
  - PEOPLE COACH THEIR PEERS, AND ACCEPT COACHING BY THEIR PEERS
  - PEOPLE ARE MORE INCLINED TO STAY, THEY ENJOY THE CULTURE.
  - FOCUS ON CAREERS, NOT JUST JOBS

OBSERVABLE BEHAVIORS

- IT IS SAFE TO TAKE ON RISK, EVEN TO FAIL SOMETIMES, IN THE FACE OF SERIOUS CHALLENGE.
  - PEOPLE ARE WILLING TO TAKE ON CHALLENGES
  - FAILURE IS NOT AUTOMATICALLY PUNISHED

Chapter 4
Navigation by Grown-Ups

When people feel that they are part of something important, when they know where they are going and what part they are expected to play in getting there, then the culture is immeasurably improved. Good navigation is a driver of good culture.

Navigation traditionally means guiding the ship along a path that safely reaches the desired destination. Our use of the term has a broader meaning; "navigation" as we use it here encompasses a range of activities:

- Determining the desired future state of the organization, product, project or service.
- Determining the plan and schedule to achieve that future state.
- Defining the intermediate goals to be met on the way to achieving that future state.
- Determining the disturbances that might occur, along with the contingencies that can be brought into play.
- Evaluating progress en-route to the next defined destination, making mid-course corrections to schedule, resources, and targets.

It's no secret that these matters are anything but trivial, and it is difficult to be perfect in all of them. Some organizations even manage to get them all wrong, so wrong that the plans and goals seem to be guided more by misplaced optimism and caprice than anything else. People in those organizations can only hope that eventually the grown-ups will show up and take charge.

This chapter discusses effective navigation and its role in driving towards a healthy culture.

A hand on the helm

It's all very well to talk about navigation being effective, but the first essential is that there has to be some navigation happening at all. That's not always the case. Some for-profit companies, for example, rely on an implicit strategy, broadly understood, that the company exists for one purpose and one purpose only: to make lots of money. But that isn't a strategy at all. Making lots of money is a side effect of achieving a vision, of growing in some explicit way at some designated time in the future. Without a clear statement of that vision, people at all levels are left to work out for themselves what the desired future state should be. That is, the steering function of this particular vessel is being done by anyone with an oar in the water. That includes you, your peers, your superiors, and your underlings; essentially everybody.

While letting everybody steer might sound democratic and egalitarian, it's not a recipe for arriving at any particular destination. You can't work in such an undirected environment for long without realizing that strategic vision needs to be communicated from above (though the inspiration for that vision may have come from anywhere in the hierarchy). The sense that someone above you has a clear and clearly communicated vision is essential for you to feel comfortable getting on with your part of the whole. This is something that everyone has to believe for the culture to perform at its best:

SHARED BELIEF *There is a hand on the helm, strong direction coming down from above.*

> *Netflix began as a company that rented DVDs through the mail. I've always been impressed that the company was not called Diskflix or Mailflix. What is clear to us all now, and was from the beginning to Netflix management, is that the mail business was only a startup phase for a company that always intended to provide video over the net as soon as the technology would support it.*
>
> *— Tom DeMarco*

If a little is good, isn't more better?

If you've ever participated in a strategic planning meeting, you might have encountered the following all-too-common phenomenon: once one strategic initiative is announced, there is an incentive to carry on and declare a whole bunch more. Coming away from the meeting with a dozen, or several dozen strategic initiatives might seem like success, but it isn't. Strategic initiatives can only work if the organization has the resources (or is willing to acquire the resources) needed to successfully focus on all the initiatives. Too many initiatives almost certainly means that few or none of them will be completed. The hard part when finding the core set of strategic initiatives is almost always deciding what not to do.

A truly strategic initiative is likely to be ambitious, and almost certainly innovative. Ambition and innovation require focus, so the ground-level workers charged with implementing an initiative should probably be concerned with no more than one.

The future imperative

At any given level – project, process, a mid-level division (e.g., a department) or enterprise as a whole – an essential shared belief is that:

SHARED BELIEF *It is the requirements of the future business environment that shape our current direction.*

Strategic initiatives aim to keep the organization relevant for the future business environment. Strategic plans made today are not about today's environment. Instead, they are trying to anticipate the environment – the customers, the habits, the technology, etc. – that will be current at the time of roll-out. If the initiative is truly strategic, it will be ambitious enough to need possibly several years before it comes to fruition. Thus, grown-up navigation is the plan that steers the organization into the future. People doing the work want (and need) to know where they are going and what target they are working toward.

Strategy and tactics

As we move down the hierarchy, navigation shifts from mostly strategic to mostly tactical. At a project level (mostly tactical), for example, the final project outcome has to be aligned with the organization's strategic direction. If it isn't, people will tend to feel underutilized or even useless. Who wants to work on a project that is doing something other than help the organization achieve its promise? A belief in the essential character of ongoing work helps a culture feel valued:

> **SHARED BELIEF** *The work we're engaged in now is consistent with the organization's strategic direction.*

An immediate implication of this is that the people doing the work understand the organization's long-term direction. Since the "long term" might be five years out, people need more immediate guidance for their day-to-day efforts. For that, the work is divided into medium-range goals, and objectives. It's these that focus workers on moving toward successful completion of the work.

You don't have to be a veteran of many such planned endeavors before you realize that it is not trivial to plan meaningful, intermediate pieces such that successful completion of each piece does indeed move the whole endeavor forward. If you've ever found yourself laboring to complete a task that seemed to you to be divorced from the real needs of the organization, or had no bearing on the long-term objectives, then you know how discouraging that can be. This is not to say that every team member has to have power of veto over any task or approach he/she feels is not needed. But it is important for people to believe in the process they're being directed to follow:

> **SHARED BELIEF** *The mid-range goals of our work plan are consistent stepping stones towards the success of our long-range plans.*

This is most obvious in project work, but also applies to any continuous process: If you're the safety officer of an assembly line, for example, your goal of containing on-the-job injuries to well below industry standards is clearly consistent with production goals. If you're the corporate lawyer, your goal of ensuring the legality of all operations is consistent with and contributes to the organization's long-term objectives.

Coping with uncertainty

This is the spot where you might expect to find a section with a heading like "Basic Estimation Competence" that would assure you that there can be no reasonable navigation without the ability to foresee how long the work will take and what budget and schedule should be applied to it. "Get your act together," the section would tell you, "and perfect your estimating skills, otherwise your plans are a joke."

But nobody has perfect estimating skills. If an endeavor is sufficiently ambitious, the unknowns can be huge. A sure way to doom any undertaking and to damage the culture of the people engaged in it is to wish the unknowns away. This form of denial is not so rare as you might hope. We know of one corporate bigwig whose people said of him, "Often wrong, but never in doubt." You know this *illusion of certainty* is doing its nasty work when you see things like this happening:

- A project reporting *Work Completed* in one status report that bears no resemblance to *Work to Be Completed* set out in the prior status report.
- A manager whose team has failed to meet a milestone still insists that the final deadline is safe.
- Work that the schedule calls for to begin by a date certain is declared to be begun even though prerequisite tasks are far from completed.

In such cases it's pointless to fault the various managers' inability to correctly forecast time needed and/or bad luck encountered. The thing to fault is that they had failed to acknowledge their own uncertainty. When this happens it's not due simply to the manager's uncertainty about project's uncertainties ("How could I have known…?"). More often it's due to denial, due to an attitude that being uncertain is somehow not allowed for real professionals. But denial doesn't make the uncertainty go away. On a two-year project with a dozen people, for example, it's almost certain that one or two of the workers will leave and have to be replaced. Is there a provision for that in the budget? Is there delay built into the schedule to offset workers' time needed for interviewing and hiring replacements? Or for being away because of illness? Are there tasks on the schedule that may not have to be done at all, that might be avoided by catching a lucky break? Or is only *good* luck built into the schedule?

> *I was involved in a project for an insurance company. The project was to build a new company almost from scratch. Naturally enough, finding the requirements for this company was complicated and time-consuming, given that expertise was needed from a significant number of people, not all of whom were in the same country.*
>
> *Time was critical, as it always is. The top manager was anxious to see some progress being made and started to pressure the project team to commit more of the project to implementation. "In fact," he announced, "I am updating the Gantt charts to show we are now in implementation."*
>
> *"You can call it what you want," responded one of the team leads, "but we know we have not yet uncovered all the crucial requirements. Until we have them all, any implementation is likely to be wrong."*
>
> *I liked this team lead. She showed that sometimes, just sometimes, sensible time management can triumph over apparent expediency.*
>
> — James Robertson

What makes people good at dealing with uncertainty is not that they are superb estimators but that they are reasonable risk managers. They build into their plans provisions to cope with the risks that their foray into the future will surely entail. They plan some tasks to be done to mitigate for risks that might never manifest themselves.

They identify their own uncertainties and try to quantify them. Nobody's perfect at this but everyone who tries is better than those who are stuck in denial. Most importantly, people doing the work have to be given reason to believe:

> **SHARED BELIEF** *The plans governing our work have made explicit and generally adequate provision for the uncertainties in the planners' assumptions.*

This is not to say that all the uncertainties have been recognized or that all the provisions for them will prove to have been adequate. But people have to have confidence that this kind of risk management is built into all the quantitative aspects of the work.

Resisting the siren call of the urgent

Uncertainty is real whether you've made provision for it or not. Team members often find themselves called away from their scheduled work to deal with urgent matters that no one anticipated. Since we can't expect to be perfect at foreseeing the future, we should expect to do a bit of firefighting. However, when there's a lot, when scheduled work is chronically interrupted, employees may become chronically reactive. What they react to is not just fire, but to anyone calling Fire! Anyone who wants immediate attention to his/her needs learns that all that's required is to assert the urgency of those needs.

Your culture should pride itself on being responsive, but not reactive. The responsive team has a clear-headed determination not to let the urgent overwhelm the important.

> **SHARED BELIEF** *Some things are urgent, some are important. Urgent does not always trump important.*

Decisions at many levels

Planning mostly happens higher on the hierarchy than the work to be performed. People doing the actual work are more likely to have a detailed understanding of the methods and techniques best used to get the work done than the planners. The plans and objectives governing the effort need to recognize this and focus on the *whats* to be accomplished rather than the *hows* of doing the work. That encourages people to be confident that:

SHARED BELIEF *Decisions are made at different levels, generally by the people with the knowledge and understanding to make them properly.*

Observable behaviors of grown-up navigation

Navigation is hard, and nobody ever gets it perfectly right from the beginning. Raising a child is hard and nobody gets that perfectly right either. The essential in either case is that those affected have confidence that the effort is thoughtful, unstinting, informed, realistic, and makes adequate provision for correcting the errors that are bound to arise.

When that confidence is present in a team or workgroup, or throughout an entire organization, certain observable behaviors manifest themselves:

People know what they are supposed to be doing

They have a navigation plan that sets out their direction for the foreseeable time. They are confident that unless some serious emergency happens, they are free to concentrate on their work and will not be constantly dragged away from it. They are aware of their objectives, and how these objectives contribute to the long-term outcome. Work is seldom subject to a "drop everything" interruption.

Prioritization is relentless

Leaders at all levels practice continuous prioritization of the tasks at hand, even when it means that sacred cows are pushed down the list.

People don't feel governed by uncontrollable events
Schedules are realistic and include thoughtful provision for the most likely risks. Success looks achievable to the people charged with making it happen.

People's time is less fragmented
Fewer fires to put out means fewer interruptions and less multi-tasking. Less fragmentation, less context switching, almost always means better, more coherent results.

No one has the sense of being over-controlled
The plans set out goals and objectives but do not prescribe the methods and techniques for doing the work. Once the goals are properly communicated and understood, the troops figure out for themselves the best way to achieve those goals.

A sense of purpose energizes the workers
People are confident their work serves the best interests of the organization. Workers make a better effort when they know their output is making a meaningful contribution.

Say it loud

Our final word on navigation is this: don't keep it a secret. People dislike working in the dark; they want to know where they are going with their work, and why they are going there. Publicizing the plan and some of its rationale is obvious and simple to do – it ensures that everybody is included in the overall scheme.

> *A colleague of mine at a government mapping organization, Samantha, was hired to improve the way projects in our organization were working. It became clear to Samantha that none of the projects had commonly understood goals. This was making it impossible for teams to communicate and respond to changes. She set about doing a goal analysis on a high priority piece of work, and came up with a consistent, measurable goal. But simply having the goal was not enough – people needed to be able to relate to the goal, and they weren't doing so.*

This lack of commitment to the goal led Samantha to write it boldly and colorfully, on an oversized piece of card. Wherever she went the goal went with her, "Uh oh, here's Samantha and 'The Goal.'" The result was that everyone became more conscious of the goal. At every project meeting The Goal had its own chair. This was amusing to people at first, but it conveyed a powerful meaning. "Is what we are saying consistent with the goal? If not, then either we are off track, or the goal needs to be changed." We were all pleased and impressed that this shared goal awareness made it possible to prioritize and make decisions, and to navigate a safe route through the project.

— *Suzanne Robertson*

Happy to Work Here – Understanding and Improving the Culture at Work

Chapter 4 Visual Summary

SHARED BELIEF
- SOME THINGS ARE IMPORTANT, SOME ARE MERELY URGENT

OBSERVABLE BEHAVIORS
- PEOPLE ARE NOT GOVERNED BY UNCONTROLLABLE EVENTS
- PEOPLE ARE NOT CONSTANTLY FIRE FIGHTING
- RELENTLESS AND CONTINUOUS PRIORITIZATION OF THE INITIATIVES

THERE IS A HAND ON THE HELM, STRONG DIRECTION COMING DOWN FROM ABOVE.

IT IS THE REQUIREMENTS FOR OUR FUTURE BUSINESS THAT SHAPE OUR CURRENT DIRECTION

THE WORK WE ARE ENGAGED IN NOW IS CONSISTENT WITH THE ORGANIZATION'S STRATEGIC DIRECTION

PEOPLE KNOW WHAT THEY ARE MEANT TO BE DOING.

ACCEPTANCE OF CHANGE AND MUTATIONS.

ACCEPTANCE OF UNCERTAINTY

Chapter 5
The Bong of Collective Confidence

"Bong" in this chapter title was originally a typo, but we thought it too good to correct.

The *bond* of collective confidence is not just a cause of good culture, but also partly an effect of the prior four drivers. Once it begins to manifest itself it becomes an important driver of further cultural improvement. In that collective confidence is a hybrid of cause and effect, it is different in kind from the other drivers. A second difference is that the bond is only indirectly actionable. There is nothing you or anyone else can do to *make* collective confidence happen, though there are things you can do to make it possible.

Collective confidence is different from self-confidence

You can imagine a team of individuals who are all supremely self-confident. This is a good thing, but not at all what we mean by collective confidence. Collective confidence arises from a shared understanding of co-workers, their strengths and needs, and a sure sense of their dependability. It derives from the dawning awareness that the other members of the team are competent individuals, and that the team as a whole can be confident that it is capable of great things.

The Dave Brubeck Quartet was one of the best jazz combos in the middle of last century. Dave Brubeck played with complex time signatures and this was offset by Paul Desmond's cool, smooth saxophone playing. From time to time Desmond would break away from Dave's melody and improvise his own piece, all the while being backed by the rhythm section of Gene Wright on bass and Joe Morello on drums. All of them were sublime musicians – they would not have been in this band if they weren't. Each trusted the others enough to go with them when they went on their improvisation. Joe Morello's drum solos are worth seeking out if you need an example.

So here we have collective confidence: each supremely talented musician willing to go on his own, or to follow one of the others, or to play together as a harmonious group. Each trusting the others to make good decisions, and each being trusted to do the same.

Each one of them knew that he was talented. They didn't need to be told. When you're a musician at that level and playing in a band such as the Brubeck Quartet, you just know you're good. Additionally, each one knew that they were part of something big.

You can't *instill* this kind of confidence; it is a direct result of successful performance by the team, performance that belongs to them all rather than to any subset.

As time goes by, the team does more and better work. It takes on more difficult projects. Team members work hard to come up with new ideas, and make the best use of each other's talents. As they succeed with each difficult project, it gives them the confidence to try something a little bigger and a little more difficult. They succeed again, and again, and every success builds their collective confidence.

Along the way, a bond forms among the team members. They see themselves naturally as individuals, but also as a part of something bigger – they develop a bond of confidence in their collective ability to successfully take on almost any kind of challenge.

SHARED BELIEF *Team members believe that, when acting collectively, they can achieve anything within reason.*

When people in a team are collectively confident, they become unafraid of bigger and riskier challenges. And so, we see collectively confident teams face up to (within reason) riskier, and thus more rewarding, undertakings.

SHARED BELIEF

Team members share a belief that they can be bold and, within reason, take risks.

Most modern work teams are made up of people from different disciplines, often working in different places. But this does not stop the paralegal admiring the work of the web designer, nor the business analyst thinking that the branding guy is doing some great stuff. All members of the workgroup know what is required of them, and each is confident that the others in the group will deliver their part.

Over time, as they work together, they learn more about each other. At a superficial level, they gain understanding about each other's skills and knowledge. But their understanding goes much further – they also become more familiar with each other's character. They see how the others behave under stress; they see how the others react when things are going well. Finally, they come to learn something about the others' personal lives. Before long, team members have developed a rich awareness of each other and of the team as a whole.

One of my team leads managed an example of collective confidence in action. The project was to replace a collection of older business information systems with a modern integrated suite of software products. At the time this happened, the project team was entering its endgame in preparation for the launch of the new suite. Unexpectedly, the business organization came up with a new requirement: historical data from the old system needed to be accessible to enable occasional efforts to study past business results. Suddenly, the team was faced with a substantial new effort introduced late in the project.

Once the business ruled that the new system could not go live without the ability to view the older data, the project team responded calmly. A new companion project was launched – very informally – to define, design, build, and test a separate archive system. This would enable access to all the unconverted data stored by the old information system.

> *The team's confidence in its abilities determined the way it reacted to the new requirement. Once the business need was clear, there were no complaints about sudden scope creep. The project was on track to meet the planned go-live date and so it was decided not to delay it. The team was experienced enough to know that it was not practical at this point to add fresh resources to take on the new requirement. So, they just got on with it and did it.*
>
> — *Steve McMenamin*

Uncertainty

In the presence of uncertainty, the confident team and its management make decisions as soon as responsibly possible. This does not equate to "Shoot first, ask questions later," but neither does it mean taking on seemingly endless study and analysis until risk is entirely eliminated. Remember that this is a team with a healthy attitude about time (Driver 2).

"As soon as responsibly possible" for the confident teams means that there are still some unknowns, but self-belief says that it is the time to go ahead. If things go wrong, then the team is confident that it can fix whatever the problem is. In fact, because there is still some uncertainty, there is an expectation that some things will go wrong. But for now, the team makes the decision to get on with it.

SHARED BELIEF *Workers at all levels believe that their collective ability will prevail despite the uncertainties that might lie ahead.*

Hygiene

Promoting collective confidence is the responsibility of leaders at all levels, but it isn't exactly leadership – it's more like hygiene. This kind of hygiene is something that leaders should be practicing regularly and as a matter of course. They cannot assume that collective confidence will happen automatically.

There are four essential kinds of hygiene:

Keep teams together: Stability of the social unit is necessary for any healthy kind of collective confidence to emerge. Workgroups need to be kept together for long enough for people to become familiar with each other's abilities. People need to know who is dependable, who can provide technical help, and who needs help.

Invest: If, indeed, people and teams are all important (Driver 1), it only makes sense that the organization invest in them. This includes clerical and technical support, training, satisfactory workspace, and tools.

Afford opportunities to work closely together: Collective confidence gets a boost when the people solve tough, complex problems together. The bonding that results from conquering a difficult situation becomes an asset in everything that follows.

Encourage: While the members of the Brubeck Quartet knew they were good, the applause they got at the end helped. When the crowd at Carnegie Hall rose to its feet, it made the players feel extra good. They would not have felt good if they had played badly, but when they knew they played well, and the audience was heartedly signaling its appreciation, it felt extra special. Collectively competent teams know how good they are. They know the work they've done is well done. They feel it collectively and individually. Even so, it helps if from time to time, to be told that their work is appreciated.

This kind of hygiene is essential for a satisfying work experience. It makes possible the bonding that lets teams become excellent instead of just good.

Observable behaviors

When a bond of collective confidence is driving your culture, most or all of the following behaviors are likely to show up:

Increased willingness to take on difficult work
The more confident the group is of its competence, the more it seeks out challenging projects.

Increased willingness to take on scope
The confident culture is prepared to make larger commitments.

Shared appetite for the work and shared enjoyment from it
Doing good work is a heady experience. Doing it with other good people is even better.

> *The book* Peopleware *that Tim Lister and I co-wrote was about teamwork and partnership. We approached the writing as a team and the resultant book was the product of a partnership. I had known the satisfaction before that of having a book of my own published and being well-received. But to my surprise the pleasure of creating half of a good work was greater than the pleasure of creating it all. A lot greater.*
>
> *— Tom DeMarco*

Increased tolerance for uncertainty
A team with strong and mature collective confidence is willing to take on projects that are more complex and more innovative. These projects bring with them greater uncertainty. The team is confident that whatever problems are encountered, they can be overcome.

Tolerance for imperfectly defined projects
Many projects start out with vague, indistinct objectives. Collective confidence allows a team to take on this kind of work, knowing that the collective skills of the members will enable them to clarify project ambitions and deliver the right result.

Greater adaptability
Sometimes people drop out of the team – illness, vacation, urgent family obligations, and so on. However, when a team possesses collective confidence and trust, it works to heal itself. The members are familiar enough with the work and skills of their

colleagues to adapt to the new circumstance. The team has the capacity to effectively cover for a missing teammate.

Self-propelled, self-correcting teams

As much as one whole level of management can be effectively absorbed into the team. For example:

> *I was consulting on a multi-year development project at a utility company. The project was several years into it, with about a year and a half to go. I spent most of my time working with S_, the project manager, and with his direct reports, six team leaders. On a Friday afternoon in January S_ went home early, feeling very ill. By Sunday he was in intensive care in a major hospital, fighting for his life. On Monday morning the team leads gathered with S_'s boss to hear the dreadful news. The boss promised to keep everyone posted on S_'s condition and asked if everyone knew what they needed to do on the project, and if there were any questions about the project and the plan, please come to him. No one ever went to him with questions; they talked with him about progress, and decisions they had made. No one was ever named Interim Project Manager. The project moved forward. S_'s desk was the way he left it on that Friday afternoon. The team leads knew what to do, and they did it.*
>
> *About five or six months later S_ came to the office for the first time, just for an hour. The team leads gathered around him, and S_ asked them how the project was going. They answered in unison, "Fine."*
>
> —Tim Lister

Finally, an essential role of leadership is to ensure that the team's collective confidence develops toward modesty rather than hubris. There is no magic to doing this. It simply requires leaders to model the behavior they are looking for. Be the behavior you seek.

Happy to Work Here – Understanding and Improving the Culture at Work

Chapter 5 Visual Summary

OBSERVABLE BEHAVIOR

INCREASED WILLINGNESS TO TAKE ON MORE DIFFICULT, CHALLENGING WORK

TEAM MEMBERS BELIEVE THAT THEY CAN BE BOLD AND, WITHIN REASON, TAKE SOME RISKS

SHARED BELIEF

TEAM MEMBERS BELIEVE THAT, WHEN ACTING COLLECTIVELY, THEY CAN ACHIEVE ALMOST ANYTHING

INCREASED WILLINGNESS TO MAKE LARGER COMMITMENTS

SHARED APPETITE FOR THE WORK, AND SHARED ENJOYMENT FROM IT

RICH AWARENESS OF EACH OTHER AND THE TEAM AS A WHOLE.

INCREASED TOLERANCE FOR UNCERTAINTY

TOLERANCE FOR POORLY-DEFINED PROJECTS

WORKERS AT ALL LEVELS BELIEVE THAT THEIR COLLECTIVE ABILITY WILL PREVAIL DESPITE THE UNCERTAINTIES THAT LIE AHEAD

Chapter 6
The Perceived Value of Excellence

Jiro Ono is a sushi chef who started his restaurant, Sukiyabashi Jiro, in the Ginza district of Tokyo back in 1965. Despite humble beginnings, Jiro has now been awarded three Michelin stars, the highest accolade in the restaurant business. (If you are not familiar with three Michelin stars, think Pulitzer Prize, a best-picture Oscar, being knighted, winning the Monte Carlo Grand Prix, or being awarded the Légion d'honneur.) How can sushi be good enough to win three Michelin stars? By being infinitely better than the stuff you have at the corner place for lunch; by being as good as it is possible to be. This is what Jiro set out to achieve; he wanted to be the best, to make the best. He succeeded – his product is so good that people have to wait months for a booking and pay significantly for the privilege of eating Jiro's sushi.

The cult of excellence

Jiro Ono is not alone. There are many people, and many organizations, with a desire to bring about the best product or perform the best service in their class. There are also organizations engaged in some ongoing process – insurance claims, air traffic control, vehicle registration – who desire to make their process efficient, effective and productive. In other words, excellent.

To become excellent, you have to believe in excellence.

SHARED BELIEF *The organization's products, services and solutions are excellent – the best that can be produced at a sustainable cost. The organization's offerings are the best available to customers in their price range.*

There is a palpable desire in the organization to achieve the highest quality in everything it does. Quality that brings in sales is just good business practice. Customers like excellent products and continue to buy them; companies that produce excellent products rarely go out of business. Some take excellence a further step:

> *The Procreate drawing app for the iPad is a product I'd have bought happily (at ten times the price) even if it were considerably less elegant. It comes with a level of quality that is way beyond what is required for commercial success. It causes you to make involuntary soft moaning sounds of appreciation. The only explanation I can think of is that the level of quality was set by the makers for their own satisfaction rather than by the budget.*
>
> — *Tom DeMarco*

Organizations that work to achieve excellence find that there is a significant return. The most notable is that excellence attracts excellence. When an organization is turning out excellent products, excellent people want to work there. If you are the best law firm in Manhattan, the best lawyers want to work for you and not some storefront outfit in the suburbs. If you are an excellent architectural practice designing the best and most celebrated buildings, top graduates are going to want to be part of your firm.

Excellent employees are certainly more expensive than mediocre employees, but excellent people are far more effective and efficient. Two of this book's authors carried out an exercise with computer programmers from all over the world and found a staggering 25-to-1 differential in productivity. The elite of these programmers was certainly not paid 25 times more than the worst, but whatever they were paid, it was a terrific investment for the employers.

Culturally, it is good to have excellent people work for you. Excellent people are disinclined to engage in office politics and troublesome behavior – that's for the B-list people. Excellent people just want to get on with doing excellent work. From a cultural as well as a commercial point of view, to invest in delivering the highest quality products and services is one of the most rewarding, and actually one of the safest, investments an organization can make.

It is in the work itself where you observe excellence. People want their products and services to be as good as they can possibly be: you never hear anyone say, "good enough." People are obsessive about the quality, usefulness and usability of their products. They almost never settle for the first proposed solution but push on looking for better ones. This often means that three, four or five proposals are rejected before they arrive at the final version.

Testing is relentless. Excellent people desire to release a product or service which is flawless, and also liked by its customers. Products are not released until – within reason – the crew judge that they cannot be made any better.

Excellent companies produce only excellence.

Organizations that strive to be excellent places to work

Organizations known for excellent product and service quality tend to be outstanding places to work. There is a belief in these organizations that employees are valuable and so they do everything they can to make the workplace a humane place to work, even when the workplace is a distributed collection of locations in different time zones. These organizations treat employees as treasured human resources and expect them to work hard and do their best possible work.

Indeed, it is hard to imagine that anyone can do excellent work unless the organization is making sufficient effort to assure that the workplace supports them. The Pharaohs built the pyramids using slave labor, but that approach is not only illegal today, it is guaranteed not to work. People want to feel valued and should be valued. Providing an excellent atmosphere to work in is a significant contributor to excellent work.

An example of an organization that produced excellent products and was an excellent place to work was Hewlett-Packard in its golden days. The company made a wide range of scientific and industrial measuring and other computing devices, many of them ground-breaking. The founders, William Hewlett and David Packard (Bill and Dave tossed a coin to determine whose name came first), developed a management style that people started calling "the HP Way." In Hewlett's words, the HP Way is:

> "A core ideology, which includes a deep respect for the individual, a dedication to affordable quality and reliability, a commitment to community responsibility, and a view that the company exists to make technical contributions for the advancement and welfare of humanity."

When Bill Hewlett gave up the helm at HP, he was interviewed at his retirement gala. He was asked what he was most proud of among his accomplishments at HP.

"Jobs," he said without hesitation. "Great jobs for people to make good money doing fascinating work, creating no pollution, making no weapons." Creating jobs, he said, that's what he was most proud of.

"Oh. What about the computers and instruments that HP builds?"

"Those are nice, too," he answered, "mostly because they have allowed us to create some 30,000 good jobs."

Companies that operate in some variation of the HP Way usually have a similar attitude toward jobs. A worker comes to believe that the organization is concerned that each person's job be more than just work for pay; it should be a meaningful experience. Creating jobs, and making them good jobs, is part of the organization's culture.

SHARED BELIEF *One of the organization's roles is to provide good, meaningful jobs and take special care of its employees.*

Adobe has achieved its goal of pay parity between women and men in the United States, India, and almost 40 other countries. Pay parity for Adobe means ensuring that employees in the same job and location are paid fairly to one another, regardless of their gender or ethnicity. Adobe is not the only company that has achieved, or is on the way to achieving, pay parity; many other organizations see it as part of their corporate ethos.

More than money

> *"Young people don't want to work for a company if it is seen as harmful to the environment or society, they want to be proud to say where they work."* — *Jaideep Prabhu, Cambridge University's Judge Business School.*

Many modern organizations believe that they exist not just to produce goods and services, but also to fulfill civic and environmental responsibilities. They are aware of the impact that their products and services have on the human population and the ecosystem: they do whatever they can to reduce any negative impact. These organizations usually undertake extracurricular activities and provide company resources to benefit the community and the environment.

Company-sponsored volunteerism is often seen in excellent organizations. This usually involves employees being given paid time off for their volunteer efforts (80% of Civic 50 companies allow employees paid time off to volunteer), or receiving "dollars for doers" grants, or the company matches employees' contributions to charitable causes, or in some other way facilitates employees to participate in community engagement activities. Microsoft sponsors *Hack for Good*, which applies the skills of volunteer Microsoft employees to pressing societal problems. Some organizations include volunteering in employee performance reviews. The benefit to both is that volunteering sometimes provides an opportunity for employees to demonstrate skills or aptitudes that otherwise might not be seen by managers.

Most of the volunteering is done by employees wanting to volunteer for their own satisfaction and wanting to contribute to their communities. The organization's part in this is actively supporting the volunteers' efforts. IBM has a program called *On Demand Community*. This program helps employees and retirees find volunteer activities that match their skills and expertise. Since the program's inception, volunteers have logged more than 20 million service hours.

Community engagements almost always start as idealistic efforts – the organization and its employees wish to do good. But something else happens. Employees who volunteer for community engagement, or some other effort, score higher in job satisfaction ratings. They are also more likely to stay longer with the organization. Companies that have a positive social/environmental impact – a little over 30% of American enterprises – find that such initiatives boost employee engagement, encourage them to stay with the organization. Hasbro, a multinational toy and board game company, found that its employees rated the community engagement program as second only to year-round half-day Fridays as the reason why they liked to work there. Additionally, companies with active community engagement programs find it easier to attract new talent.

> *"Society is increasingly looking to companies, both public and private, to address pressing social and economic issues. These issues range from protecting the environment to retirement to gender and racial equality."* — Larry Fink, BlackRock

On a smaller but nonetheless impressive scale, TOMS Shoes LLC is very actively doing good. The founder, Blake Mycoskie, wanted to help disadvantaged children to have shoes. Shoes for a Better Tomorrow gives a pair of shoes to a child in need for every pair sold. TOMS so far has given away 86 million pairs of new shoes. TOMS also donates eye care for each pair of sunglasses sold, as well as having active engagement with charities working for clean water and safe births.

People, especially young people, want to know that they are making a positive contribution to humanity or the planet (which is pretty much the same thing).

SHARED BELIEF *It is good to do good.*

Self-actualization, or realizing full potential

Think of an organization sitting on the top tier of Maslow's hierarchy of needs. In case it is too many years since you read Abraham Maslow, his hierarchy of human needs builds from the bottom up with *physiological* needs – food, water, shelter, etc. – then *safety*. When you achieve personal safety, your next need is *belonging and love*. When that is achieved you climb the hierarchy to achieve *esteem* – respect and recognition from peers, and finally you make it on to the top step, *self-actualization*, where people realize their maximum personal potential. This progression is illustrated in Figure 6.1.

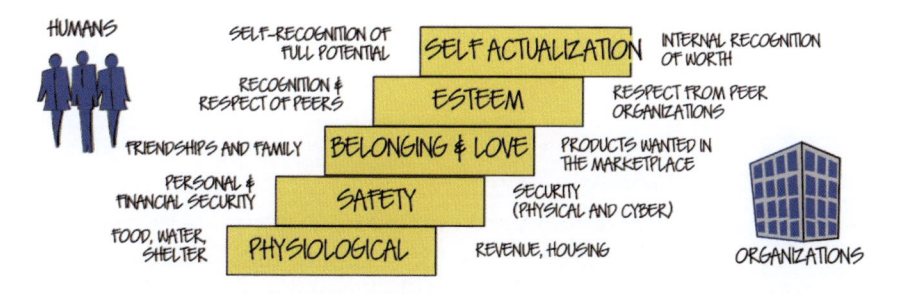

Figure 6.1. Maslow's Hierarchy of Needs. Humans start on the bottom step and progress up the hierarchy as they achieve each level of need. Note that you have to achieve one level before thinking about moving up. Not surprisingly, organizations follow a similar path.

It is easy enough to understand Maslow's hierarchy from a human perspective, but what about that of an organization? We start with revenue on the bottom level. Without revenue the organization cannot function, just as a human cannot function without food and water. Once the organization has revenue and housing, it needs the security of rule of law, and to be a safe place to work. Safety means that it is able to protect itself from physical and virtual threats. Then it moves up a step and produces products that are valued in the marketplace.

Thereafter, excellent organizations progress if they are respected by their respective industries. Other organizations have a lot of respect for, or want to be like, companies such as Apple, The New

York Times, Federal Express, Amazon, Hewlett-Packard in its former incarnation, the UK's Special Air Service (SAS), Netflix, and Nike. There are many, many more, but you get the point.

To get to the self-actualization step, the organization must honestly think of itself as being as good as it can possibly be. This is not just motivational posters put up by a well-intentioned HR person. It's more than a boastful CEO proclaiming without any real justification that his company is world-class. This is about *knowing* that the organization – to paraphrase Maslow – has become everything it is capable of becoming.

SHARED BELIEF *Excellence and beneficence matter. They matter a lot.*

Observable behaviors in organizations that value excellence and beneficence

The underlying belief of management and employees is that they are doing all they reasonably can to be excellent and beneficent. The following behaviors can be observed by anyone who cares to look for them:

Engagement

There is an evident satisfaction in the work and in being a part of an excellent organization. Voluntary resignations are much lower than the industry norm, and there is a widespread sense of belonging within the workplace.

Transparency

Good organizations are willing for their customers and employees to know what the company is doing, what their development plans are (within reason), what problems they are having, and what they are doing to address these problems.

Craftsmanship

Most of us look upon the work as an achievement in itself, but some – and here Japanese artisans are standout examples – see each accomplishment as one small step along the way to perfection. This, even when the artisan feels that he or she will probably never achieve that perfection. It is not the achievement that matters so much as the straining to reach the perfect state. This behavior is characteristic of cultures of excellence where you never hear anybody say, "Good enough."

Shared ethic of design

Design is everything and everything is designed. Design is part of the normal culture for good organizations. That is, everything is designed for maximum effectiveness and functionality. (It can also be pleasing to look at, but that is not the point.) The workplace is laid out so that employees can properly interact as a work team. Processes, products, and services are designed to provide optimal value, rather than for cheapness of manufacture or execution. Painstaking, obsessive care is taken to ensure that what the product does is done in the best, most convenient and logical manner. It is an unwritten rule that nothing is shipped until everybody is satisfied that it can't be any better. And design is not skin deep – it goes all the way down. The inside of the product is as well designed as the outside.

> *"If you don't pay attention to beauty, then you are consigned to a world of expediency."* — Unknown author

Community involvement

Excellent companies often have a strong sense of belonging to the wider community around them – they recognize their social obligations. They encourage staff members to volunteer to put their skills to work for the community (often on company time) and provide direct financial contributions and/or donations of products to complement their work.

Environmental responsibility

Excellent organizations are aware of the environment and their contributions toward it, and what they may be doing to harm it. They are aware of the impact of their emissions, packaging, transportation, energy use, waste, etc., and are doing whatever they can to reduce that impact. The organizational culture sets an example that employees are expected and encouraged to follow.

No organization can be perfect, but it can drive the culture to be as good as it can be. If your organization is to be excellent and/or beneficent, preferably both, then its shared belief system must include a strong sense of what is right and what is wrong. It must be constantly trying to come onto the right side of the ledger. Being and doing good has nothing to do with mission statements and corporate reports. It has everything to do with what the organization believes and does.

Chapter 6 Visual Summary

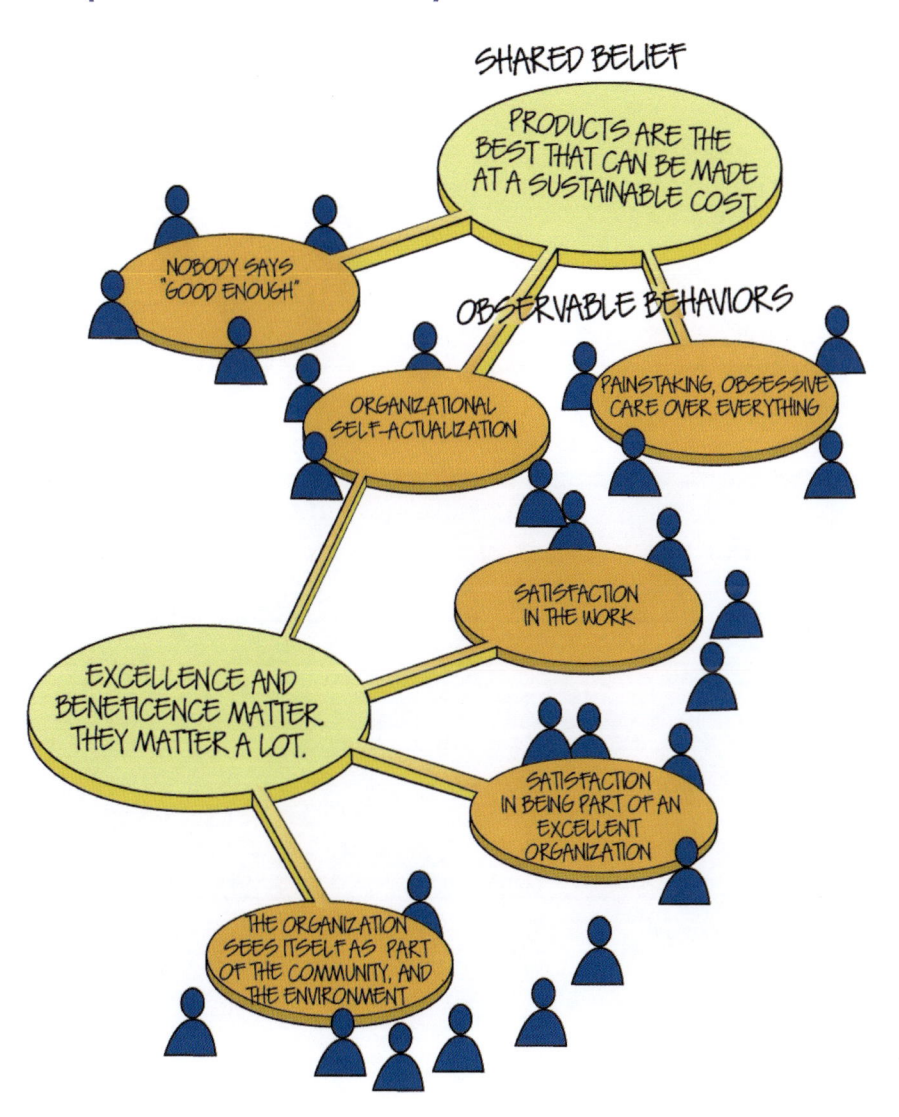

SHARED BELIEF

- PRODUCTS ARE THE BEST THAT CAN BE MADE AT A SUSTAINABLE COST
- NOBODY SAYS "GOOD ENOUGH"
- EXCELLENCE AND BENEFICENCE MATTER. THEY MATTER A LOT.

OBSERVABLE BEHAVIORS

- PAINSTAKING, OBSESSIVE CARE OVER EVERYTHING
- ORGANIZATIONAL SELF-ACTUALIZATION
- SATISFACTION IN THE WORK
- SATISFACTION IN BEING PART OF AN EXCELLENT ORGANIZATION
- THE ORGANIZATION SEES ITSELF AS PART OF THE COMMUNITY, AND THE ENVIRONMENT

Culture Drivers Redux

"Happy families are all alike; every unhappy family is unhappy in its own way."

— Leo Tolstoy (opening line of Anna Karenina)

All happy organizations are alike in many respects. Think about it backwards: You would hardly consider your workplace culture a happy and healthy one if you didn't feel safe there, or if your skills and talents weren't appreciated, or if you had no clue what any of the goals were, or your time wasn't being used wisely, or there was no pride of workmanship, or if your team wasn't confident it could rise to challenges. In all our combined experience, we have yet to encounter a truly healthy culture without some evidence of positive shared beliefs in all six areas. They are the essentials of good culture.

[Diagram: Six culture drivers surrounding "WORKING CULTURE" — Excellence, People, Time, Safety, Collective Confidence, Navigation]

That doesn't mean that the shared beliefs are positive all of the time. People have off days and lose focus, tempers flare, managers sometimes let their optimism get the better of them and sign on to hopeless schedules. The important thing is not that the organization be perfect; the important thing is that all or most of the shared beliefs underlying the drivers are core to the organization's value system and everybody knows it.

Unhealthy (unhappy) cultures can be ailing in a great number of different ways. These are the subject of Part II, just ahead.

Part II
Culture Killers

Part I was about factors that drive good culture, and now Part II is about something else entirely. It explores the ways that culture can be undermined and ruined by norms that evolve without anyone's explicit permission. We've identified twenty-eight of these culture killers, each one the subject of its own short chapter.

Taken all together, these culture killers make up a true chamber of horrors. As you read through them you may find yourself exclaiming that surely no company or organization anywhere could be so stupid as do all these culture-destroying things. And you'd be right. But you're likely to find some chapters that describe something eerily familiar to you from your own work experience.

While no organization is guilty of all of them, each one of the Part II essays describes something we've encountered somewhere in our practice. Each of them is real someplace, though most of them, thankfully, won't apply where you work.

Unspoken rules

Culture killers often take the form of rules that are broadly understood and obeyed within the culture, but almost never spoken out loud. For example, some organizations have the unspoken rule, "Don't announce that a key project milestone is going to be missed till you know pretty precisely how much more time is needed." Applying this unspoken rule means that if your project is clearly going to be late, you defer telling the bad news (maybe even for a few months) while you figure out a new safe date. But this rule has some terrible consequences: people who have a right to know about schedule slip are kept in the dark and the project manager and team are living a lie during the interim. The rule in its baldest form is "Lie about schedule status till you're ready to reschedule." Bad for the organization as a whole, but also particularly bad for team culture.

Most unspoken rules are unspoken because they are, well, unspeakable. They are not recorded anywhere – we would be embarrassed or ashamed of them if they were, or we would be forced to repeal or revise them. Nobody speaks them out loud for the same reason. These are the bad guys. These are the rules of behavior that result in a toxic culture – the culture that you do *not* want.

While these unspoken rules can do terrible damage to a culture, they are paradoxically among the easiest things to fix. Each one is a clear indicator of actionable culture improvement. Once you identify a toxic rule, repealing it can be as simple as bringing it into the light of day. When you say the rule out loud, the damage it can do will be readily apparent.

In this part of the book, we lay out a collection of toxic unspoken rules. If you find one that applies where you work, you'll know what to do about it: Say it out loud. Set out to repeal or revise it. Use the identified toxic situation as a diagnostic tool to steer you toward which of the healthy culture drivers needs your particular attention.

Chapter 7
Internal Competition

Most organizations are in competition with other organizations. Corporations strive against each other. Even in government, different departments compete for funds.

If competition is healthy – if competing well is a goal – doesn't it stand to reason that setting people *within* an organization to compete against each other is healthy, too? In organizations where such internal competition is encouraged, the unspoken rule seems to be:

Internal competition makes us fitter.

The question is: fitter for what? It certainly makes us fit to compete ever more fiercely with our peers. It's something of a stretch, however, to assume that competing internally somehow makes us fitter to compete in the broader market.

Jack Welch, formerly CEO of General Electric, became a darling of Wall Street by setting his people at all levels to compete with each other for favor. He did this by the simple expedient of firing the lowest performing 10% of his workforce every year. Yes, it made employees miserable, but investors thought the scheme was just swell. They drove the stock price up . . . at least for a while. But General Electric, almost from the moment that the whole angry business started, began to lose ground in the marketplace and eventually the stock followed. The once proud company that Welch inherited has never been the same since.

The lifeblood of an effective culture is cooperation, and internal competition is hardly conducive to cooperation. After all, the person *you* cooperate with might well use your help to enhance his/her position over you, to your detriment. Competition doesn't have to be zero-sum, but it often works out that way.

A sad victim of zero-sum competition within an organization is peer coaching. (Why would you ever help someone who was being schooled to compete against you to your detriment?) But coaching matters a lot; losing it diminishes us all. Think of a situation in your own career when a peer took you aside to help you improve a key skill or enhance an understanding that would help you work better. Felt good, right? And it probably felt just as good to be the helper in cases where you coached others. We look back with gratitude at times when our peers took the time and effort to coach us. Thank you, Coach. Thanks, too, to the organization, since it was the organization that somehow made such coaching opportunities safe for everyone: safe for you to acknowledge need and safe for your coach to help you without fear of eventually coming to regret it.

The unspoken rule that encourages internal competition is damaging to the culture and works to the detriment of everyone in it. If this is the rule in your workplace, you need to repeal it.

Chapter 8
Followership

The word *team* is sometimes erroneously defined as "a group of people who follow the orders of one leader." Leadership, in this view, depends on obedient followership from those below. The unspoken rule here is:

> **Teamwork means doing what you're told.**

There are many words frequently used to describe healthy organizational cultures, but "obedient" is not one of them. The authoritarian notion that "I manage the team and the team members do what I tell them to do" is painfully wrong for teams of knowledge workers. It makes the manager the sole source of strategic and sometimes even tactical thinking. It damages the culture and reduces the team's ability to work effectively by stifling its initiative. But we see this all the time. Part of the reason is our first-family upbringing where parents lead, and children are expected to follow obediently. That model may feel *familiar* when applied in the workplace, but it is nonetheless ill-suited.

A variant on the same theme is the unspoken rule:

> **There are no failures of leadership, only failures of followership.**

An unspoken rule that encourages followership is likely to be the death of true leadership. And it damages the culture.

Chapter 9
Politics

In a work context, the word "politics" usually refers to activities that have little to do with effective operation of business processes or getting product out the door, but a lot to do with the enhancement of an individual's perceived importance. It can also mean that person A is trying to score political points over person B, or make person B look bad. It can also mean that person C is trying to enlarge his or her sphere of influence or increase the number of subordinates.

You probably have other meanings for the word, depending on your experience of politics in the workplace. Whatever meaning you attach to "politics," it will have nothing to do with improving the organization, its effectiveness, or its products.

Unfortunately, politics can play a role in almost any company. Unfortunately, because such politics are rarely beneficial, and for some bizarre reason, the intensity of the politicking is often in inverse proportion to the importance of the issue. Which brings us to our unspoken rule:

Political advantage is more important than getting useful stuff done.

Chapter 10
Hierarchy and Network

In the pre-modern world, large organizations were rigorously hierarchical. Think armies and the church. If you occupy any level of such an organization, everyone directly beneath you is under your command. Power is exercised directly down the hierarchy. Communication moves down the hierarchy, not up. When you command someone to do something, he may respond "Yes Sir," or "Yes, Your Grace," but since that was the only possible response, it conveys precisely zero information.

Peers on such a hierarchy are not under your command. They have domains of their own, removed from your influence. They may be competitors. They are certainly competing with you for the move up to the next level: there is only one spot there directly above, so it's a zero-sum game who gets it. Now imagine that one of your subordinates is communicating directly with one of the subordinates of your competing peer. Erghhh. This is threatening stuff. No good can come of it.

Modern organizations are not so strictly hierarchical, but not so different either. There has to be networking across lines of the org chart in order for the operation to function at all, but it can sometimes be a bit threatening. You may even have encountered situations where workers in one manager's domain are discouraged from interacting directly with anyone in a peer domain. All interaction has to go up the hierarchy, through the boss, and then down. The unspoken rule at work when this happens seems to be:

Networking is subversive.

This is a really dumb rule because we all know that networking is absolutely vital; that people who are good at it are star performers. The rule can only be a rule as long as it remains unspoken. But don't think for a moment that just because it's dumb that it may not be alive and well in some organizations, maybe even yours. The way power is granted (advancements made to a higher position on the org chart) can sometimes seem capricious to everyone. Given that, there is bound to be some insecurity felt by those in the middle levels of the org chart. The more insecurity, the more likely it is that the unspoken rule of subversive networking will govern.

Chapter 11
Slack

You might expect that the more time-focused your culture is, the more it will abhor any kind of slack. Slack is time that is not strictly allocated to a task that is essential to reaching a declared goal. From the viewpoint of an old-guard industrial manager, this looks like inefficiency. In fact, such managers were diligent about driving slack out of any operation. At least from their perspective, the unspoken rule was:

Slack is waste.

And yet, the best organizations, particularly those doing creative work, thrive on a certain amount of slack. It's during slack time that reinvention happens. A bit of slack is also necessary to be quick in responding to direct customer requests (everyone is not so busy that the customer has to wait for attention). And finally, it is during slack time that the culture is discussed and improved and healed. If everyone is busy 100% of the time, by definition there is no time left for culture. There is no time left for anything.

The most effective operations are not the busiest ones. That seems like a contradiction in terms, but it isn't.

Chapter 12
Overhead

If you've been in the workforce for more than a few years, you've probably noticed that non-professional support people are being cut everywhere. Gone are secretaries, clerks, typists, along with many lab technicians, librarians, editors, researchers, graphic artists, and in-house tech support.

In short, what's been eliminated from your workplace (compared to that of your parents or grandparents) is overhead. All this is done in the name of efficiency. The unspoken rule of such efficiency programs seems to be:

Overhead is evil.

Cutting overhead is most often justified in terms of efficiency, but it's really more about cost-reduction. And what's getting reduced is *short-term* cost.

We humans are less than perfect at trading off between short-term cost and long-term benefit. For example, a truly superb short-term cost-reduction scheme is to buy an ancient junker of a car rather than a spiffy new model with better dependability, safety and fuel efficiency. Talk about cost-reduction: on the day of your purchase, you're spending maybe $1,000 versus the $25,000 or $30,000 a new car would cost. On that day, you are a genuine cost-reduction hero. Yes, but how about the next day and the days after that? If you're aware of the downside of buying old junker cars, you can see where this is going.

Every time support people are removed, the tasks they used to do are picked up by the people they used to do those tasks for. Work once done by a low-paid clerk, for example, is now done by a much more highly paid scientist, doctor, manager, or engineer. This is hardly a recipe for efficiency. In fact, this kind of efficiency "improvement" makes the organization less efficient. More important, it makes the organization a lot less effective.

Grapple for a moment with us over the key question here: how does an organization allow its net efficiency to be *reduced* by an efficiency improvement process? Like any good detective, you have to ask yourself, Cui bono? Who benefits? Not the organization as a whole, of course, but someone in the organization does benefit. Cost reduced from payroll goes directly to the bottom line. And the person whose efforts have thus padded the bottom line may well gain power and possibly a fat bonus.

When leaders of countries sacrifice the well-being of their countries to increase their power and line their purses, we call that *corruption*. Gaining power by cutting support so that professional people have to do low-level tasks is corruption, too.

If this is how an efficiency program works in your organization, maybe you need an anti-corruption program.

Chapter 13
What Doesn't Just Happen

Some organizations react to any setback by assessing blame. The unspoken rule in such cultures seems to be:

Shit doesn't just happen; it is somebody's fault.

In a blaming culture, it is not a matter of what went wrong, but who caused it. It is not possible to have a mishap without a culprit. A blaming culture is unpleasant to work in, but that's not the only reason to avoid it. A far more serious reason is that when blame is automatically assessed, the natural tendency for self-critical appraisal is thwarted. And self-critical appraisal is perhaps our most powerful way of learning.

When a healthy culture team has failed in some way or other, someone is sure to start the conversation about why it happened with a line like, "Okay, where did we go wrong?" When you hear that, you can bet that learning is about to happen. If the answers begin to sound familiar, then a second question is warranted: "Why didn't we learn from the last time this happened?"

Missing a goal might be a catastrophe, but it should never feel like a disgrace. The only disgrace would be not to learn from it.

Chapter 14
Staring into the Future

How do you respond to this?

"You started out thinking we had four months' worth of work to get this finished. After the first month you needed an extra week to get where you needed to be at month's end. Same thing happened in month two: you needed an extra week. So, now you are two and a half months in. Please give me an estimate of when you will deliver."

If you say, "Unless we change something, we're looking at five months," then you are showing perfectly reasonable, adult behavior.

If you say, "We can recover. We'll be done in four months," your unspoken rule is:

Never admit trouble until there is no remaining way to pretend it won't happen.

If you say, "We're over the rough patch. We'll be done in four and a half months," your unspoken rule is:

Be optimistic! Past problems will never happen in the future.

Chapter 15
Always on Top of It

There is a name for an annoying adult who hogs the conversation with dubious facts and stalwart opinions. Such a person is called a "know-it-all."

There are some organizations that appear to be entirely populated with know-it-alls. Listen in on meetings, on groups trying to make a decision. How did the organization hire 100% know-it-alls?

It didn't. It *created* them by establishing the perverse rule that admitting you don't know something is admitting weakness. You betray weakness by admitting you are not on top of it, no matter what *it* is. The unspoken rule is:

Never say "I don't know," even if it is true.

If you feel that you lose status by saying "I don't know," recognize that you are living a fraud. Healthy organizations are full of I-don't-know-admitters. Telling the truth is okay, really.

Chapter 16
The Business of Busyness

An old joke to set the tone for this section:

> *A group of excited young curates crashes into the office of the Archbishop at St. Patrick's Cathedral. "Your Eminence!" one of them cries. "Jesus Christ has just appeared in lower Manhattan!"*
>
> *"What?!"*
>
> *"He walked across the water and came ashore in Battery Park."*
>
> *"Oh my."*
>
> *"And now he's headed up Fifth Avenue toward St. Patrick's. He could be here any minute!"*
>
> *"I see."*
>
> *"So, tell us, Eminence, what do we do?"*
>
> *The Archbishop thinks that over for a moment and finally says, "Look busy."*

An apparent busyness can be a sign of deep and very professional engagement in an important task, vital to the long-term interests of the organization. Or it might be a sign of something else entirely. In a fearful organization it most likely implies a worry that it's downright unsafe to seem unbusy. The unspoken rule that governs people in this case is:

Look busy.

Of course, the fear itself has already done damage to the organization's culture. But obedience to the unspoken rule makes the matter worse. The consequences of everyone trying to look busy include:

- No time for reflection
- No time to confer with colleagues (which might be interpreted as "chatting")
- No time for lunch
- No time for training
- Nobody willing to be away from his/her desk
- No off-site activities
- A general uneasiness with activities that might seem "passive," like reading and research

Most of the things that the rule makes impossible are culture positive. That is, they help the culture heal and improve itself. The more you find yourself and your co-workers compelled to look busy, the surer you can be that your working culture is damaged.

Chapter 17
Knowledge is Power

Knowledge is power says the well-worn cliché. Knowledge is often hard won; it may take several workgroup months to uncover some fact, some technique, some technicality, that can rightfully be considered a breakthrough.

So, what does the workgroup do with the newfound knowledge? Does it rush to share the knowledge with the rest of the organization as usually happens in the scientific community? Or does the workgroup hide it away and decline to share it with others?

You may have observed the phenomenon in your organization where two (or more) workgroups are working on similar projects, but there is little or no sharing of information. Why? Enter the unspoken rule. If we know something that they don't, doesn't that make us more powerful? Even if it is not to the common good, it is to *our* good to keep our knowledge to ourselves. The unspoken rule is:

Keep hard-won knowledge to yourself; it makes you more powerful.

Chapter 18
The Anti-Innovation Society

"We don't do things that way around here."

"I don't think management will go for it."

"Someone tried that before."

"No, that would be too hard."

"That's a completely untested idea."

"Things are fine as they are; they don't need to be changed."

Anti-innovation organizations are painful to work in. No matter how good your idea is, no matter how beneficial it would be, someone, somehow, finds a way not to make use of it.

This is not necessarily a management issue. Of course, a blaming culture, or one where any failure, no matter how small, is punished, discourages innovation and risk-taking. Similarly, when your workmates are disparaging about any advance or new idea, it sucks the life out of your creative thinking. Any change or improvement is resisted, even when the change would be beneficial to the resistors.

In short, any fresh thinking whatsoever breaks the unspoken rule:

Innovations? We don't need no stinkin' innovations.

Chapter 19
Kick the Can Down the Road

When confronted with the need to make a difficult decision, the organization decides to defer the decision. Sometimes this is done by commissioning another study (never mind that several studies have already been done), and sometimes by conjuring up some reason why the decision can be delayed ("We need to wait to see whether interest rates will go up.")

Kicking cans can become an art form. It means that people don't have to do anything because, having kicked the can down the road, they now have the perfect excuse for inaction.

The unspoken rule is:

Whenever possible, delay making any difficult decision.

The hope is that when the time comes that a decision can no longer be avoided, it's become somebody else's problem.

Chapter 20
Denial

One problem that can be the death of any strategic planning exercise is people's propensity to deny that a potential unfortunate event could ever happen. They continue to deny despite evidence that suggests it could well happen. Sometimes history has strong lessons to be learned about the future, but those too are denied:

The virus will fade away when the weather warms up.

Oil will never again go below $50 a barrel.

The Wall Street banks are too big to fail.

People won't stop buying our product.

Digital photography will never replace film.

The Red Sox will never win the World Series.

People won't want to post things about themselves on the Internet.

"There's no chance that the iPhone is going to get any significant market share. No chance." — Microsoft CEO Steve Ballmer, 2007.

What unspoken rule is at work here?

Denial is easier than facing uncertainty.

Chapter 21
Authority

The word "authority" has two quite different meanings. If you are *an authority* on a subject, that means you know quite a lot about that subject. And if you are *in authority* over a group of people, that means you have positional power to manage and direct them.

Sometimes people in authority conclude that their positional power automatically makes them an authority on anything and everything that happens in their domain. The unspoken rule here is:

Positional authority confers subject matter authority.

This may have been a pretty fair assumption in organizations a century ago. This was when managers' knowledge and understanding of all the work going on beneath them was the very thing that propelled them into management. Today this is a much more dubious proposition. Modern teams often depend on bringing together many different disciplines, and even the best manager can't be master of all of them. Managers of such cross-functional teams have to be ready to push some decision-making downward. That is, they decide who makes the decisions rather than make the decisions themselves.

Following the unspoken rule assures that decisions get made too high in the organizational structure and risk being therefore less connected to the real criteria that ought to drive the decision.

Power corrupts when it makes a powerful person feel almost omniscient. Consider some of the autocratic (or wannabee autocratic) national leaders now in power in various countries. It doesn't matter which one you pick, he is bound to be someone who ignores expert counsel from advisors and subordinates, and makes each decision based on his own inclination. This is clearly not a recipe for the best decisions. When dissenting voices are silenced (fired, demoted, jailed, or accused of disloyalty), then the very fact that they aren't heard anymore may be judged sufficient confirmation that the boss really does know best.

Chapter 22
Overtime

Extended overtime is a culture killer. A little is fine, but after weeks and months of putting work ahead of family and home life and evening and weekend leisure, it begins to feel abusive. Teams become fragmented since the overtime load is unevenly distributed. What may have started out as an enthusiastic response to challenge begins to feel like being used. A sense of being used doesn't drive people together; it's more likely to drive them apart.

Companies that are most effective at extracting overtime from their workers have somehow managed to establish the following unspoken rule:

A willingness to work overtime proves that you're a professional.

While most professional workers are willing to put in some occasional overtime, that doesn't mean they are unaware of abuse when it happens. The blithe assumption of overtime in making up budgets and schedules is an abuse. It treats the people who will have to do the work as chumps.

It also sends the message that the work is not very important; if it were important, the budget and schedule would have made adequate provision for it without depending on contributed overtime. When the budget and schedule are stingy, it will take more than pep talks to convince people that what they're doing really matters.

Chapter 23
Brownie Points

One of my most depressing training experiences happened when one of my students upon entering the room, went immediately to the back of the classroom. I always ask everybody why they are there and what do they want from the course. One student told me that he had no interest at all in the course; he was attending because his manager told him to be there. It turned out that the compensation scheme at this organization awarded managers brownie points for training their staff, but it didn't matter whether the training was applicable to the job or not.

— *James Robertson*

This student had inadvertently revealed one of his organization's unspoken rules:

Pretend to be interested in your staff's personal development; it can be rewarding to you.

Chapter 24
Anger

A pattern of anger in the workplace is about as damaging to a healthy culture as anything we can think of.

Of course, there are sometimes policies and blunders and stupidities that would anger any reasonable person, but in those cases it's the policies and blunders and stupidities that are hurting the culture, and the anger is simply a side-effect.

But now consider an anger that is itself the problem. Consider the possibility of someone in power exhibiting anger because *that person believes that anger can be a tool.*

But a tool of what, you might ask. Chances are your experience has already suggested one likely answer. There is no question that an angry boss can raise the stress level for any direct report. The short-term effect of raised stress is increased attention. If your boss yells at you, you are going to be suddenly very much in the moment. If that's what the boss wants, the anger has served its purpose.

The unspoken rule such a boss is following is:

Self-righteous anger is a good way to convey urgency.

A powerful person's anger will get your attention *for a while*. It may make you scurry out of the room, seemingly about to get cracking on whatever it is that person really wants you to get cracking on. But the long-term effect is quite the opposite: it just makes you want to thwart anyone inclined to act like such a bully.

Remember this yourself if ever you're about to explode in anger at a subordinate. It may work to good effect once, but not twice. It does you no good in the long term, and it harms the culture for everyone else.

Chapter 25
Cliché Culture

Ben (not his real name) goes to see his boss. He tells his boss that the new quota system is forcing unhealthy shortcutting of some of the key steps in the group's work. This has resulted in missed or misdirected shipments, and more congestion in the transport hubs.

Ben's boss responds:

> *"Ben, let me be clear about this: our number-one priority here at Suchard Logistics (not its real name) is customer satisfaction. It is our mission to fulfill the dreams of the people who ship with us, and our passion is happy customers. We are also 100% dedicated to providing shareholder value and the bottom line is part of the sacred journey. We pride ourselves on the efficiency and effectiveness of our carefully crafted business solutions. They are there to make our employees happy employees. I think I have answered your question. If there is anything else, anything at all, my door is always open."*

Why is the cliché such a culture killer? Two reasons: first is that clichés cover up incompetence or ignorance. The second is that the cliché is mostly a lie: "Our thoughts and prayers" is a lie. "We are passionate about" is a lie. "Win-win situation" is a lie. "Let me be clear about this" is a lie.

Clichés and good culture rarely are found in the same place because the cliché culture is governed by this culture-damaging unspoken rule:

Clichés are great: you can use them to avoid answering awkward questions.

Chapter 26
The Cafeteria

You may have visited a company cafeteria (perhaps it was your own) before or after the lunch rush, and found two, three, or four people around one of the empty tables discussing some work-related matter. Several tables might be similarly occupied, and you would have noticed that they are well spaced apart.

What's going on here? The answer is simple: people have come to the cafeteria because there is nowhere else in the building that is quiet enough to have a sane meeting, or their desks are so close together that they cannot meet without disturbing their uninvolved workmates.

In such companies, management has an unspoken rule:

Why bother with proper workspace when we have empty tables in the cafeteria?

Chapter 27
Hot Desking

Hot desking is the process of employees erasing all traces of their existence at the end of each working day, putting any possessions into a locker, and then the following day, reclaiming their possessions and finding a desk to work at. This process is repeated daily. This approach to real estate has accountants happy about the amount of expensive city office space saved.

Hot desking may sometimes be justified. A firm of consultants, for example, would want its consultants to spend most, preferably all, of their time racking up billable hours at the client's office. We must also allow for organizations that make use of rented shared workspace.

But consider the permanent on-site employees. The message hot desking sends them is that they are simply not valuable enough for the organization to give them their own desks.

They must, each morning, find, not be given, a place to work. This apparently takes on average 18 minutes. And then, when they do find a desk, there is an even bigger problem: since there is no adjacency of co-workers, they have almost no direct contact with any other human. Everything is done by email or message. You might well be asking why these people are in the office at all. If they wear headphones and have nothing but electronic contact with co-workers, then they would be better off working from home.

The unspoken rule for hot desking organizations is:

Real estate outranks employees.

Chapter 28
A Day Without Lunch

If you've been employed for more than about fifteen years, you have probably observed the same change we've seen in attitudes toward lunch. Before this time, lunch with workmates was so common that it was essentially a norm. And since? Since then, companies have bifurcated, some making a huge fuss over the value of lunch with co-workers, while others have seen the phenomenon almost disappear.

First the fuss-makers: You don't have to work for a high-flier company to become aware that some of them have elaborate cafeterias, multiple serving stations with different menu options: Chinese, Thai, barbecue, made-to-order sandwich and burger stations, carveries, and elaborate desserts. Some have free food and optional table service, outdoor lunch spots, and espresso bars. Do an internet search for "best companies for lunch" to see some mouth-watering possibilities, and not just in Silicon Valley. The message these companies are sending is that lunch with your colleagues is clearly something to be encouraged, that it's completely consistent with the organizations' goals.

The other side of the coin is organizations where lunch is most times taken alone at a desk. The cover story for this varies, but the result is the same. Sometimes people say they are too busy to take a lunch break; sometimes people feel guilty if they take a lunch break and their colleagues don't; some people feel that doing menial tasks while eating lunch at their desks is the same as taking a break.

Whatever the thinking, the result is people eat at their desks and keep on working. This despite laws in most countries that mandate that employees take a break, and despite most managers understanding the benefit of employees taking a break with their colleagues. And yet they don't, even in organizations that aren't busy at all.

The unspoken rule in these companies is:

Lunch with colleagues is a luxury to be indulged in only when time permits (and time *never* permits).

What's saddest about this is that lunch with colleagues is a culture-building experience. It's over lunch that you learn that Arlene's son has just won a scholarship, that Larry has a quirky sense of humor, or that others (not just you) are uneasy about the coming expansion plan. If there's a difficult relationship with a peer organization, you learn who the players are and how to deal with each one. None of this happens in a lunchless culture. Whatever the willingness to trust was before lunch, it may be higher after. Teams that take lunch together are much more likely to be trusting and familial.

To end on an optimistic note: the trend toward lunchlessness has been successfully resisted in both France and Italy.

Chapter 29
Standard Operating Procedure

Somebody other than you knows better how to do your work.

Sometimes work procedures are designed by "specialists" who don't themselves do the work requiring the procedures, and in the worst case, have never seen the work being done. This often results in the people doing the work finding that the procedure does not fit the task – an insult to the people who know better but were left out.

Chapter 30
Gyrating Like the Price of Bitcoin

An unspoken rule we've encountered far too often:

Only what is urgent today matters. Yesterday's urgency is not so urgent anymore.

I was seconded to a larger branch of my company in another country. Apart from the enjoyment of visiting a new city (for me) for a couple of months, I found some interesting work practices that we did not have at home.

One of these was the weekly meeting with the managers of a business section, where they set down their priorities for the work that the support sections were to do for them. I initially thought that this was cool, until a few weeks into the process I noticed that the priorities changed each week. Something that was super-hot one week would be replaced by a different super-hot item the following week, and resources pulled away from last week's priority item.

Nobody finished anything. Everybody was switching from one task to another, and the organization had almost zero continuity. This didn't affect me too badly – I was there to do something that did not fall under the auspices of the fluctuating business section – but colleagues around me were decidedly unhappy. Why shouldn't they be? They were rarely able to experience the satisfaction of finishing something, and more or less began each task knowing they probably wouldn't finish it.

Was anybody doing long-term strategic planning? If they were, I failed to notice it. Was the inconsistent re-prioritization playing havoc with the culture? My colleagues certainly thought so.

— *James Robertson*

Chapter 31
Bureaucracy Is Its Own Reward

A recent change in regulations governing French *copropriété* (the legal entity that administers a collective of separately owned properties such as a condominium) means that every *copropriété* must fill out and submit a complicated and time-consuming form each year. The French Republic has existed for quite a long time and has not seen any need for an annual form to be filled in. But now it says it has a need.

Why should this be so? There has been no significant change in the real-estate world to justify the new document. No, this is the work of the bureaucracy, in this case it is the French Civil Service (*Fonction Publique Française*) – the huge, cumbersome and opaque bureaucracy that carries out the government's business.

The French annual *copropriété* form happened because someone in the vast empire of the Civil Service (the word "civil" is used from its origin of relating to ordinary citizens; it should not be mistaken as meaning "polite") wanted to expand a little, and make more work, so *voila!*, tens of thousands of *copropriété* forms being filled out all over France, and hundreds more civil servants needed to process them.

Bureaucracies love to grow, which they do by creating more (sometimes completely unnecessary) work. For many of the people in a bureaucracy, the ultimate goal of their work is not to get product out the door or provide a service, but to make more work. As the work they know is bureaucracy work, they make more of that.

The same kinds of bureaucracy exist in many private companies, usually larger organizations, and parts of them that have become so far removed from the prime service or product that they have, for all intents and purposes, lost sight of the organization's reason for existence. For most of the people in it, the bureaucracy is their universe, so understandably they desire to make it larger and more complex. Both result in more bureaucratic and administrative work:

"Let's start monitoring the percentage of parking spaces being used by the hour."

In addition to more work, we have the bureaucrat's favorite extra benefit – it makes the administration more complicated:

"We can have everyone eating in the canteen fill in a food allergy card, add a bar code to that, and if they go near a dish on the buffet that has their allergy, sound an alarm. Of course, the cards will have to be renewed each month."

Any additional work is good for the bureaucracy.

Bureaucratic overreach reveals an unfortunately prevalent unspoken rule:

Never make anything simple when an unnecessarily complicated alternative can be constructed.

Chapter 32
Anger, War, and Chaos

The following is an extreme example of a culture gone awry. We've only encountered this one time, but it's nonetheless instructive. If you see anything like this in your company, seek employment elsewhere.

> *I went to a new client's main office for the very first time. I expected to spend most of my time with my contact, but he told me that we would be attending the CIO's 9:30 meeting. He then opened a desk drawer and pulled out a bottle of vodka, about 2/3 full. He asked if I wanted some, but I declined. He poured himself a generous shot and slugged it back. Off we went to the meeting. The CIO was joined by her six subordinates.*
>
> *The meeting was...unforgettable. When one person talked the others felt the need to comment aloud or pay no attention at all. It appeared that the CIO was just fine with this. Each person gave a progress report which was greeted by laughs, and rude – as in terribly vulgar – comments. Then the group turned to the topic of the day, which was whether to adopt a new calendar application that one manager wanted to make mandatory for everyone in the organization. Now all hell broke loose: screaming, shouting, fist-pounding.*
>
> *The CIO eventually screamed the pandemonium to a halt, turned to me, and said, "What do you make of this?"*
>
> *I said, "I don't know what to make of it. I have never seen anything like this. I have no context."*
>
> *As my vodka-swilling contact and I left the conference room, he stopped us in the hall, and whispered something to me.*
>
> *He was willing to confide in an outsider never to return, and had to whisper because it was an unspoken rule.*
>
> — *Tim Lister*

The rule at this company was:

If you see something good, grab it, and if you can't grab it, kill it.

Chapter 33
Bullshit Jobs

David Graeber, an anthropologist, speculated that many people went to work each day and did nothing. That is, nothing that was of any benefit to their organization, the world or themselves; a completely empty job that had no meaningful output. Graeber ran a survey on YouGov and 37% of people in Britain responded that yes, they had such a bullshit job. A similar survey in the Netherlands resulted in 40% saying the same. Estimates for the US have been similar or even higher.

Keep in mind that the people responding were willing to say that the job *that paid their salary* was bullshit. It is easy to speculate that many more would not admit that their job was meaningless, even if it was: "It's very important that I monitor the hourly usage of training room A."

You might be thinking that this could only apply to the public sector, but no, bullshit jobs are split pretty evenly across public and private. They appear in almost all industries – it turns out that even the highest tech organizations have their share of pointless work. You must have come across people in your own experience who either do so little, or so meaningless a task, that their job qualifies as bullshit.

Personal assistants reported that they have nothing to do, but their boss tells them to look busy – he wants to keep the prestige of having two PAs. Some assistants reported that they do the real work; their boss has the bullshit job of going to lunch and attending meaningless meetings. Many bullshit jobs are in administration. Process workers on the factory floor have been systematically made more and more efficient so that most of *their* work is real. This indicates that contrary to the popular notion, bullshit floats upwards.

Some of the unspoken rules that might apply in such a culture are:

☠ **Don't tell anyone that his or her job is bullshit.**

☠ **Don't tell anyone (except David Graeber) that *your* job is bullshit.**

☠ **Don't point out to management that bullshit jobs exist in the organization. You might be speaking with someone who has one.**

Chapter 34
The Stepford Hires

The Stepford Wives is a movie (it's been made twice, actually) based on a book by Ira Levin wherein all the women of Stepford, Connecticut, are practically identical. They dress the same way, talk the same way about the same things, and behave towards their husbands in the same manner. It turns out that they are all animatronic robots. (Sorry if this spoils the movie for you.)

Some organizations, particularly successful and popular ones, turn themselves into Stepford by hiring only those people who walk and talk just like the people already in the organization.

Sometimes, and this is a bit more insidious, HR people only hire people who look like themselves. Again, the organization begins to become Stepford.

Naturally enough, if there is no beneficial disruption, no influx of fresh faces and fresh thinking, the organization becomes predictable and eventually moribund.

A Stepford organization has this unspoken rule:

Play safe and avoid variations. Hire people only if they are essentially indistinguishable from the people already working here.

Unspoken Rules Redux

What to conclude after you've made your way through our "chamber of horrors" of all the ways we've encountered that companies can damage the working culture? If none of the examples we've cited applies where you work, that's wonderful. You can have a laugh at the expense of other workplaces where such absurdities are the norm.

If on the other hand, you've found any that do apply, you've got your work cut out for you. You can't really begin improving the culture till you've undone the damage being done by the culture killer(s). State the offending unspoken rule(s) out loud and enlist your community to help you repeal or replace. This is low-hanging fruit of culture improvement because the unspoken rules are inherently absurd, and the people you need to persuade are likely to understand that.

It's possible that the notion of toxic unspoken rules will suggest something that applies in your company that we haven't seen before and was thus left out of our list. If so, in the next chapter, try your hand at writing your own toxic essay:

Chapter 35
Do it Yourself

Here's your chance to write about a culture killer you have encountered in your work life, and the unspoken rule that enabled it:

<your suitably creepy title goes here>

_____

_____

_____

_____

  The unspoken rule here seems to be:

_____

Part III
The Good, the Bad, and the Hyggelig

So far, we have brought you the good – six drivers of a healthy workplace culture – and then the bad – the unspoken (and let's face it, unspeakable) rules that are toxic to good culture. We now come to the *Hyggelig*. This is a Danish word, pronounced hoog-ly, which has no single-word English translation, but has connotations of warm, cozy, being with friends, and being in a good place to be.

Your workplace should be a good place to be. It should be a place where you are happy to come to each day, knowing that while the work itself might be challenging and demanding, the benefits, both professional and collegial, are huge.

You should expect to be happy to work at your workplace, and to feel good about yourself and your workmates. We may not be able to do anything to help with the crappy furniture or dreary uniformity of office space, but we can do something about the culture that makes or breaks the workplace equilibrium.

In Part III – this time a single extended chapter – we look at how you and your workmates can have an impact on the culture, and how you can move your workplace towards *hyggelig*, a place where all are inclined to think and say:

Happy to work here.

Chapter 36
Effecting Culture Change

You might think of your workplace as interconnected individuals and groups, with work products, documents, ideas, critiques, emails, etc. flowing among them. The culture of your workplace doesn't explicitly alter any of the workflows – doesn't act to redirect a status report from Christine to Gratian for example – but what it does do is "lubricate" the interaction by affecting the degree of engagement felt by both parties.

On the other hand, can you imagine an organization where this lubricant is largely missing? This would be a joyless organization where hours of workers' lives are exchanged for a paycheck, and engagement is limited to whatever is necessary to scratch out the work, and very little more.

We hope that's not a description of your workplace, but no doubt your workplace could still use some improvement.

A journey in many small steps

Your task now is to do a *partial* adjustment to improve culture in your workplace. Incrementalism is going to be your friend here. Your core method will be to make a small improvement, analyze, celebrate, and then go on to do it again.

In choosing where to apply your first efforts, consider two possible kinds of target: 1) the smallest or easiest changes to pull off; and 2) the single most valuable change, the one that could have the biggest positive effect. Both are important. Easy improvement energizes the change process and helps you to recruit an ever-larger set of willing participants, while the big change – particularly when it can be approached incrementally – is where the major payoff will be.

Without knowing characteristics of your workplace (project-focused or production, matrixed or dedicated, proportion of contractors, type of work, team structures, etc.), we can't foresee exactly how culture change might proceed for you, nor can we prescribe a specific process to follow. However, regardless of the approach you take, it will have a few common identifiable parts. These are shown in Figure 36.1.

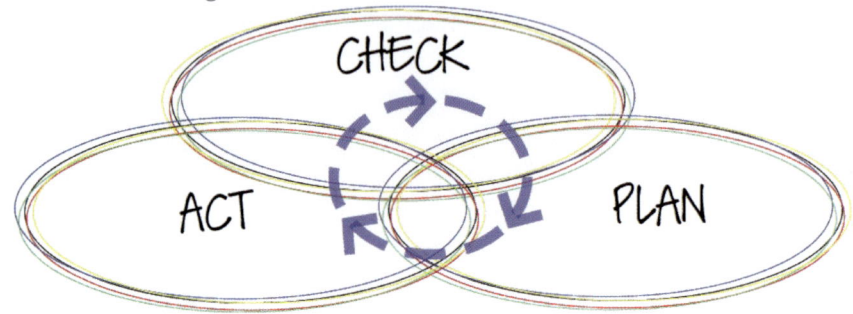

Figure 36.1. The activities of culture change.

The *Checking activity* is where you assess your current culture. You determine the extent of your cultural community, recruit friends for culture change, and assess your situation possibly by doing some of the quizzes we set out below.

During the *Planning activity*, you determine where to try culture change. Whether your change will be a more modest change or take on some of the hard stuff. You would also consider doing some risk reduction here.

Acting involves carrying out some of the suggestions we set out in this chapter. And when you have acted, it is probably worthwhile to *check* the effect of your efforts before *planning* the next move.

It's not our purpose here to set down a prescription, but rather to provide some tools and ideas that you can use to begin effecting culture change. An obvious beginning is to check your current situation and assess the kind of change – no matter how small – that would bring some noticeable and beneficial impact.

(Almost) all culture is local

If you work for a huge organization, changing the culture of the whole is beyond the possible. But most of what makes a culture healthy is determined at the local level. How you feel about your workday is very little affected by the possibly thousands of employees you don't know or interact with. We often see "enclaves" of healthy culture even in companies whose overall culture is less than admirable. The character of your culture is almost entirely a function of the people you will probably bump into during any given day, your community of interactors.

So, what is your community? In order to determine that, think of the people around you in three subsets:

- The *Everyday Circle* is made up of the people you interact with several times a day.
- The *Occasional Circle* consists of those people you go to for specialized knowledge. You consult with them occasionally, but not more frequently. They are not part of your team.
- *Beyond My Culture* contains people who are of little or no consequence to your workplace culture.

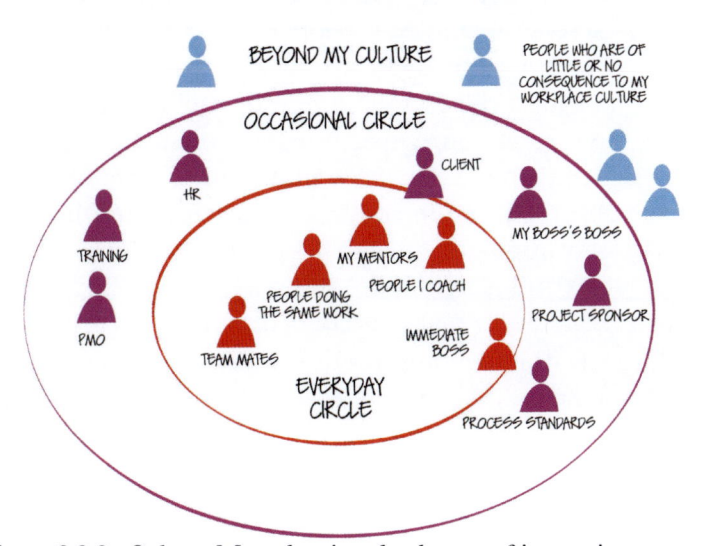

Figure 36.2. Culture Map showing the degrees of interaction you have with others in the organization. The strong and continuing connections are in the Everyday Circle.

> ***Do this****: Identify your Everyday Circle; ask others inside your circle to do the same. Merge the circles to identify your community. Make sure that no one who expects to be included is left out.*
>
> *This step is essential because it establishes meaningful bounds for the effort to follow. It's possible to improve the qualities of an identified community, but much harder to make changes that transcend community boundaries.*
>
> *You might also consider giving some kind of identity to your cultural community: the "criminal law team," or the "back table lunch group," or the group from fulfillment who call themselves the "cargo crew." Such names instill a quality of cohesion that unnamed collectives don't have. When you are looking for supporters to help you improve the working culture, it is best if your community has an identity. It is far easier, and appealing, to join an identified team than it is to join "those guys who are trying to change culture."*

Your Everyday Circle plus those additions proposed by members of that circle is what we're going to call your *cultural community*.

The exercise of culture mapping may also lead to the uncomfortable awareness that there are no real communities. It appears that everyone has Friends and Others, but nobody has a community. If this is the case in your workplace, your first step is to answer the question, "Why don't we have a community?" There can be many possible causes, but the most likely is extreme work fragmentation. Work is highly fragmented when people are multi-tasking with many different roles and assignments and may be matrixed to a specialty area that itself has no real sense of community.

Extreme fragmentation is due to poor work design, inefficient since people spend so much of their time task- and context-switching, but just because it's sloppy doesn't mean it will be a trivial matter to correct it. Start by tracking the extent of fragmentation and then campaign to reduce it over time, arguing always from the perspective of net efficiency.

Champions and activists

Change is hard and no *positive* change is likely to happen without someone to champion it. Maybe the Champion for culture improvement is you. Maybe it's your boss. Or maybe it's a rotating role with different members of the cultural community taking it on for different parts of the effort.

If the Champion is relatively powerless (here we're talking about positional power, not the power of intellect or energy), then he/she needs to enlist powerful friends. That may mean developing allies in levels above or in peer organizations.

In the best case, all members of the cultural community act as Activists. That is, they stay aware of the effort and jump in when their help is needed. They know how to grapple with specific issues for the cultural community. They have the skills to convey what's happening to those outside the community, a requirement if it becomes necessary to enlist outside help.

You can't do this alone

Since all of us are occasional windmill tilters, it's worth pointing out that you can't hope to effect culture improvement success all by yourself. The people you need to recruit to the effort are the members of your cultural community.

Your cultural community will almost certainly have at least one level of management in it. Recruiting that manager is essential because culture improvement takes time and work, effort that might otherwise be applied to doing work that that manager is accountable for. For any success to happen, there must be at least a minimal investment in people's time and energy. It may be no more than a few informal sessions or regular team lunches, but such small investments are catalytic and vital.

> **Do this:** *Recruit your boss and the rest of your cultural community to involve themselves in the work of culture improvement. Ask for their engagement.*

The management alliance

We've made the point that almost all culture is local, but without at least some management involvement your culture improvement exercise will be much more local than it needs to be. Whatever local culture you're trying to improve, that effort is best undertaken with full cooperation of the manager above it, and at least tacit acceptance one level higher.

If you are the manager of your cultural community, the following are necessary parts of your special role:

- Encourage and be cautiously willing to trust those who step up to lead. (The manager is not the de-facto leader of all culture change.)
- Work to build consensus, not to dictate change.
- The fact that you are participating at all gives *sanction* to the effort. Such sanction will begin to seem like lip-service unless you also make sure that the work is given the resources it needs to undertake improvement.
- Recruit your own boss. Be prepared to make the case to him/her of the feasibility of improvement and the differential between modest cost and likely benefit.
- Safety has to be your particular concern. People under you have to be completely confident that you will keep them safe.
- Your help will be essential if the improvement work needs to touch outside parts of the organization.

A Culture Quiz to assess health

A team exercise asking a few simple questions about workplace culture can be the beginning of the assessment discussion. Since the first three culture drivers are going to be foundational to any cultural improvement, we propose that you begin with them:

Questions about the perceived value of people and teams (Driver 1)

1. Are team members part of the process of hiring a new person who will join their team?
2. Are people hired for their talent, or are they just the cheapest labor available?
3. Are multiple levels of management involved in the hiring process? Or is it almost entirely done by Human Resources?
4. Does the hiring process strike you as comprehensive and largely focused on the work itself and methods to be used?
5. Do teams seem to be valued by levels of management above them?
6. Is there an effort to keep successful teams together?
7. Is fit of an individual to the team valued?
8. Is someone or something disturbing team harmony?
9. Do teams have any dedicated space (adjacent workspace, war rooms, etc.)?
10. Are teams being disrupted by frequent resignations?
11. Do you feel that the organization is investing in its people?
12. Do you feel valued by your organization, or are you just a necessary inconvenience?
13. Is your employee review process a frank discussion with your manager or more like a brutal, annual, impersonal points-based assessment?
14. Do you feel a sense of community with other team members?
15. Does the organization encourage or provide for societal interactions for employees? For example, are there occasional pizza lunches, or outside group activities?
16. If you are part of a distributed team or working from home, do you have frequent, that is more or less daily, societal interactions with others in the team?
17. Is there a collegial bond among peer managers?

Questions about the perceived value of time (Driver 2)

1. Is it considered a crime to waste time?
2. Is enough time allocated to do quality work while still maintaining schedules that don't feel padded?
3. Are deadlines and schedules set according to reasonable estimates, or are they set by whim?
4. Once set, are schedules allowed to be adjusted for genuinely unforeseen mishaps?
5. Is everyone always aware of the schedule?
6. Are there any activities that you consider a waste of time?
7. Are meetings conducted in a way that values your time or do they just seem to you to be time wasters?
8. Are schedules reasonable? Does finishing work on time feel like an achievable objective?
9. Is everyone aware that tasks should be finished as quickly as will allow the task to be done properly?
10. Is time treated as an overly elastic resource where projects are routinely extended rather than committing to delivering the products?

Questions about safety and security (Driver 3)

1. Do you feel safe?
2. Is the workplace characterized by mutual respect?
3. Does any kind of bullying happen in your workplace?
4. Do you sense destructive undercurrents?
5. Is it safe to tell uncomfortable truths?
6. Do you feel loyal to teammates and trust that they will be loyal to you?
7. Do you feel loyal to your organization and feel that it will look after you if times get tough?
8. Is failure punished? Or is some failure considered to be a normal part of innovating and experimenting? Is it safe to take worthwhile risks?

9. Is coaching valued? Do peers coach each other? Do peer managers coach each other?
10. Do people trust each other? Are they consistent in what they do and what they tell you?

Questions about navigation by grown-ups (Driver 4)

1. Are you aware of your organization's strategic direction?
2. If so, is the direction sharp and focused? Or scattered and vague?
3. Do you know where your organization wants to be in five years with regard to its products and services?
4. Do you and your team accept that there are uncertainties, and that your plans might have to change along the way?
5. Do your plans allow for unforeseen happenings and sheer bad luck?
6. Do you know how your work fits into the strategic direction? Or do you ever wonder why you are doing some of your work?
7. Is prioritization an ongoing activity within your workplace?
8. If you are behind schedule, what happens? Is the overall schedule adjusted? Or does everyone pretend they can catch up?
9. Do you feel that someone is in control and confidently setting and steering in the right direction? Or are you frequently at the mercy of unforeseen events?
10. Does urgent mostly take precedence over important?
11. If you compared last month's Work To Be Done list with this month's Work Completed list, would the two lists be even close? Or were you blown hopelessly off course?

Questions about the bond of collective confidence (Driver 5)

1. Do you feel that your team is capable of doing the job – the whole job – well?
2. When presented with a challenging project, do you and your teammates feel threatened or not?
3. Can you rely on all your teammates?
4. What is the attitude to risk? Something to be avoided? Or are you, within reason, willing to take some risks?
5. Is your team willing to start a project when it is known that many aspects of it are uncertain?
6. Are decisions made when needed? Or are they subject to seemingly endless analysis?
7. Are the actions of your workplace best described as bold or timid?
8. Are teams mainly kept intact from one project to another?
9. Are you regularly told by people outside your workplace that your work is good?
10. Is your team willing to take on difficult work?
11. Is the team willing to take on large-scale work?
12. Is the team eager to take on vital work regardless of how challenging it might be?
13. Does your team enjoy working together?
14. Do you confidently look forward to satisfactorily completing your current project and getting on to the next one?

Questions about the value of excellence and beneficence (Driver 6)

1. When you look at your product, can you imagine that it could be better given the price constraints of its market?
2. How good is your organization at making cost-quality tradeoffs?
3. Do people in your workplace voluntarily spend some of their own time to make the product better?

4. How busy is your complaints department?
5. Do you and your workmates take pride in what you do?
6. Do newly hired employees ever tell you that they came to work here because your product is so good?
7. Are you ever congratulated by others in the organization (including management) for particularly good pieces of work?
8. Does the organization see your work as a job or a career?
9. Does the organization exhibit a commitment to diversity, and racial and gender equality?
10. Does your workplace engage in sponsorship, charitable, community and/or environmental activities?
11. Are you given time to participate in community activities?
12. Does your organization donate part of its profits to a cause?
13. Is the organization making significant efforts to have less impact on the environment?
14. Are you happy working for the organization, and never feel embarrassed about it?

Do this: *Assess the culture of your workplace. Adopt a Culture Quiz (ours or one of your own, or a hybrid) and run it past members of your community.*

The questions about safety and security, as well as some of the others, may be troublesome for some people to answer honestly. You will need to provide a survey mechanism that shields the individual's identity to get a complete sense of what people think and how they feel about the present culture.

Open a discussion about toxic elements like the ones we've described in Part II. Are there any evident culture killers at work in your organization?

If weaknesses are turned up in any of the driver-one-through-three questions, or if there are serious culture killers identified, it's really no use

> *going on beyond. But if the first part of the analysis shows a clean bill of health, go on to assess with respect to the remaining drivers.*
>
> *Use the assessment results to identify strengths and weaknesses, as well as low-hanging fruit for improvement, and at least one area where major change could be invaluable.*

Backing into a healthy culture

Culture is important, but it's also a bit abstract. In addition to focusing the community on culture, you should also consider two variations on healthy culture: the *sense of community* in your workplace; and *loyalty among and between colleagues*. Anything you can do to improve either of these factors will also improve culture.

> *A customer accounting project that I joined was noticeable for the poor work ethic and widespread apathy. I wanted to change this and instill a better culture. I made a guess and tried something which turned out to be significant – I persuaded the team to give themselves a name. The chosen name was* CAS Cats. *Admittedly not a great name but it worked. We found a cardboard box and drew the face of a cat on it – people seemed to like having it on their desk. There were no motivational speeches, just an extremely daft sweepstake game that only the CAS Cats played. Outsiders were rebuffed if they asked to join in, and this seeming exclusiveness unconsciously strengthened our internal culture. After about six months, the CAS Cats name was more like a badge of honor, and the attitude to work had completely changed. The stronger culture led to people cooperating with each other and producing better work. Through their shared identity, people knew they were part of something special.*
>
> *– James Robertson*

Community used to be a characteristic of neighborhoods. In years gone by people stayed put for most of their lives, engaged with their neighbors, knew each other's kids and dogs, and valued a sense of community that was tied to place. Those days are gone, but people still have a need for community. For most of us who don't know our neighbors, the best chance we have to build a sense of community is in the workplace. And indeed, the best workplace cultures we encounter do feel like communities. Does yours? The things that do and could make it feel like a community are where your attention should be directed. The extent that a workplace feels like a supportive community is what makes you happy to work here.

The loyalty we referred to above is not directed to the organization as a whole, but to the members of your community. Psychological studies of soldiers going into battle tell us that no one putting his life on the line is very much motivated at that moment by patriotism (loyalty to the larger whole). Instead, what motivates them is loyalty to their mates. They move forward, even into danger because of that loyalty, because their fellow members of their small community are moving forward and possibly counting on them for support.

If you now look back over the six culture drivers from Part I, you will see that each of the drivers tends to enhance the sense of community, and each allows and encourages a growing and pervasive loyalty. Contrarily the toxic unspoken rules we described in Part II tend to have the opposite effect.

> **Do this:** *Assess your workplace again, this time from the perspectives of sense of community and loyalty. Again, use an anonymous survey mechanism to solicit true feelings. Add the results into your prior findings. Make an informal record of all the observations, and present it to the community. Make sure all members of the community have felt heard and are generally in sync. Now discuss what you can do to improve the culture within your community.*

Preserve and celebrate what's best

Before diving into improvement proper, take a moment to consider how to preserve what's already good about your culture. Holding on to your present good qualities is an important goal.

> ***Do this:*** *Memorialize (name and celebrate) positive elements of the present culture. Build awareness within the culture community of the value of these positives. Make sure that the steps you take to improve culture do not jeopardize present-day cultural strengths.*

The power of lunch

Workplace culture is a social construct, and a natural mechanism to address it is through an inherently social interaction, specifically, lunch. Regularly having lunch with members of your community is not only useful for thinking about improvement, it is also a small *demonstration of improvement in action.* The sense of community gets better over lunch, much as eating together can be a regular bonding experience for families. Loyalties tend to grow over lunch or at least be affirmed.

You could imagine that the strength of your connection to teammates is all about respect for their abilities and competence. However, equally competent robots – nameless, faceless, genderless, and featureless – would hardly inspire you to connect. What inspires connection is the person behind all that ability and competence. It's over lunch that you learn that Horace has two kids, one of them a champion diver, and that Estelle can do mimics that crack everyone up. This is not exactly work; it is culture-building that makes work more friction-free and enjoyable.

In addition, culture improvement is going to require some investment by the organization and by the individuals in it, and lunch could be that investment.

> ***Do this:*** *Build regular lunch get-togethers with your community into the life of your workplace. An enjoyable lunch is inherently agenda-less, but the discussion that ensues will often be about the culture, its strengths and weaknesses, and what can be done to make it healthier.*
>
> *Particularly in organization where most people have been inclined to take lunch at their desks, the switch to lunch together with teammates sends a powerful message. It is likely to energize the group and make clear its*

determinate to improve the culture. Additionally, it will almost certainly improve what you eat for lunch.

When part of the team is working at a distance that's all the more reason to get the visitors out to lunch on the occasions when they are united with the rest of the group.

Hard-learned lessons

We humans sometimes learn from doing the right thing, seeing that it works out well, and then continuing to do it. But more often we learn from doing it wrong the first time (or first few times) (or first hundred times) and eventually getting it right. The following are some hard-learned lessons about culture improvement, either from our own experience or that of companies we have observed and counseled:

- *Own the process.* You want to change the culture, so you have to do it. Change happens from the inside. HR can be an invaluable resource in effecting cultural change, but it's important that the process not be owned by HR or by any other outside group.
- *Start off local.* Focus first on what you can do within your own cultural community. Any broader change needs to be deferred until there is some demonstrable success to point to.
- *Apply your efforts mostly at the team level.* Don't push everything upstairs to the boss. That will be a temptation because those higher on the hierarchy have more power, but most of what you're trying to build at this point is consensus rather that a dictated result.
- *Don't waste face-to-face time.* When part of your community is working remotely or mostly remotely, treat actual face-to-face time as *the* key opportunity for culture work. Too many organizations feel it necessary to dedicate rare group together-time to structured meetings. Think *unstructured* time instead.

- *Be explicit.* Progress is going to depend on not being coy about the difficult things that need to be said.
- *Network, network, network.* Treat your improvement community as a network, not a hierarchy. Nobody is boss.
- *Go public.* Don't hide what you're doing; be willing to share successes and even a few failures with the wider world.
- *Don't over-plan.* Trust your gut about what needs change and what can be changed and how to go about it. Other aspects of your work may be at least somewhat mechanistic; this one isn't.
- *Spread the work around.* It's good to have a cultural champion, but that person should not do all the work. Keep everyone involved.
- *Think outside the company.* Consider taking on some community or environmental activity, either individually or with members of your cultural circle. Note impact on the group of doing something beneficial for others.
- *Communicate.* Articulate the (true) belief that you're encouraging others to share. Time spent working out the right way to express an idea is time well-spent. This is true in most aspects of your work, but still more true when the task at hand is culture improvement.

A bold first step

New endeavors often have some honeymoon capital to invest. Assuming your culture improvement effort has that, how should it be applied? The ideal would be something that feels big yet doesn't threaten anyone above you, and that can be viewed as a symbol of culture improvement. For example, if team members are not already given a major role in the hiring process, asking for that role could be your bold first step. Hiring is hard and uses up valuable time up and down the hierarchy, so volunteering the team to take part has the genial side effect of off-loading the managers who otherwise would bear most of the burden.

> **Do this:** *Go back over the six cultural drivers in Part I. Do any of these – People, Time, Safety, Confidence, Navigation, Goodness – suggest areas where you might start your improvement campaign? There must be some aspect of your workplace that is crying out for attention.*

Whatever first step you choose, you're going to have to sell it to people who may be willing to listen but who also may need some convincing. As you go about this convincing, be prepared to make the case not from your own perspective, or from that of the corporation (or society) as a whole, but rather from the perspective of your listener.

Your listener is one, perhaps all, of the people in your everyday circle. These are the people who have to make some change if their culture is to improve. However, people do not change what they are doing just because you ask them to. They need to believe that it is in their own best interest before they even begin to think about making a change. Up to you to paint a picture of what the workplace culture might become if the everyday circle gets behind your proposed first step.

It goes without saying that listening to your listener is going to be as important as anything he/she hears from you.

> **Do this:** *Pick your bold first step and begin painting the picture of what the culture could be for the people who need to be made part of change.*

Measuring success

One possible measurement of success is whether people stick around. Obviously, they stick around if they like working there. The U.S. Bureau of Labor Statistics has it that the median job tenure is a little over four years. You may be able to find a better statistic, one specifically tied to your industry. You can then say that if people are voluntarily and frequently leaving in significant numbers, then possibly poor workplace culture is driving them away. Of course, there are reasons other than culture that cause people to leave their jobs, but within your cultural community you should have some notion why people are departing. Given the high cost of recruitment and replacement, lowering the voluntary departure rate is an almost immediate benefit of your culture improvement.

Given that it is impossible to have complete statistical confirmation about your improvement efforts, subjective assessment by members of your community will have to serve. Query your community as changes are put in place. Key this query to the six drivers. Specifically ask whether the changes make you feel:

1. that people are now more valued, rather than still treated as dispensable nuisances.
2. that time is allocated so that quality work can be done, that time is not wasted, and never used as a weapon.
3. safer at work, or is bullying still part of the culture?
4. more confident of your workplace's ability to meet challenges and deal with uncertainty.
5. that you understand where your project is going, and what it takes to get there.
6. that your workplace is producing the best work it is capable of, and that you are engaging in community activities sponsored by your organization.

Do this: *Check in regularly with community members to be sure that changes implemented are having the desired effect. Use an anonymous mechanism to elicit even uncomfortable conclusions.*

Celebrate successes in culture improvement, even small ones.

Iterate, iterate, iterate

One thing that feels particularly good about good cultures is the sense that their improvement is never done. People in these organizations talk about their culture, talk about others they've encountered, and about ideals. The business of culture improvement is continuous. The sense that it matters as much as methods and skills and technology is ubiquitous.

Whatever success you've had we hope will be an impetus to go on to seek further success. A dedication to iteration is what makes small improvement admirable; if it all had to be done in one go, then only humongous change would be worthwhile. A small change for the better, followed by another and another and another, is a pathway to a much-improved culture.

At the same time, consider again the activities portrayed in Figure 36.1. The check-plan-act cycle is helpful when figuring out what steps you need for your next iteration.

> **Do this:** *Identify your next small advance and start moving toward it. New initiatives may require that you redefine your community; enlarge it enough to encompass the area where the benefit of change will accrue. That means you may have to reach out to people outside of your original domain and forge new relationships.*
>
> *People added to your larger community may already feel part of another community; that doesn't mean they may not also be part of yours. The places where communities intersect are areas where friction is likely to occur but also areas rife for culture improvement. The cure for friction is lubrication.*

Take your protein pill and put your helmet on

Culture improvement is a daunting but doable task. You spend a considerable amount of your life in your workplace, it logically follows that your workplace is somewhere you like to be. We cannot help you to make your work more exciting or challenging – you might have to change jobs to achieve that. But whatever work you do, we can help you to make it more enjoyable with the tools and suggestions we have made throughout this book.

The wonderful cultures we observe are not due to happenstance; they are shaped and implemented by thoughtful people pretty much like yourself.

Thanks for reading . . .

Thank you so much for sticking with us all the way to the end. Books these days live or die as a function of reader reviews. Do us a favor, please, and post a review of this book at Amazon or wherever else you acquired it. Thank you in advance.

Memories of Steve

This book is the outcome of the determination of a remarkable person, our colleague Steve McMenamin. James and I went to see Steve in California at his sheltered accommodation. He was very ill, we expected him to be resting and taking it easy. We were pleased and surprised that upon our arrival, he was ready to go to lunch at a local winery and discuss how we could collaborate on the new project he had been thinking about. That project is this book – what is workplace culture and how can it be improved?

Steve's many years of experience, early on as a developer, and later as a project manager, CIO, vice-president, and all the time as an author, made him aware of the cultural differences that exist between groups of people (small and large) and how they work together. His insight was that there are six major drivers that influence cultural behavior and create shared beliefs. He explained that if we can make those drivers comprehensible, we could build a language for understanding our culture and for influencing improvements. We all had enthusiasm for the idea and thought that if we could pool our experience, we could build a valuable guide to workplace culture. With Steve's guidance and regular conference calls we started writing this book.

Part of the writing process is listening to one's fellow authors. Steve could listen like no other human I know. This is not just hearing which we all do but listening as only Steve could do it. While struggling to explain a complex situation, I knew that Steve was listening. Never once did I feel he was wanting to interrupt or had zoned out and was waiting for me to get it over with. When I finished, his questions and comments were those that proved he had not only been listening, but had insights into the situation that could only come from wrapping a significant intellect around the problem. This ability helped to steer our collaboration and to solidify our ideas and understanding of workplace culture.

On November 20, 2019, Steve McMenamin succumbed to multiple medical afflictions. We are thankful that we had the opportunity to work with Steve on this, his last project. We miss him very much and hope this book will do what he wanted – to make it possible to understand and improve the way that people work together.

— Suzanne Robertson

About the Atlantic Systems Guild

Craft Guilds formed in medieval London when craftsmen came together to protect and advance their trade. The Guild of Barbers (which included surgeons and dentists), the Guild of Cordwainers, the Worshipful Guild of Goldsmiths (which still exists), and many others, all sought to improve their trade and become involved in civic duties. Guild members were often appointed to influential positions in the community; the chief spokesman of the Guilds became the Mayor of London.

Fast forward a few hundred years and the authors of this book formed the Atlantic Systems Guild with similar objectives, but without any aspirations for one of us to become Mayor of London. Guild principals have independent but parallel careers practicing, teaching, and improving the fields of system requirements definition, team leadership, and project management.

We principals, originally six but now sadly five, share information and our talents; we come together to work on projects – such as this book – but plow our own furrows as appropriate.

One incentive for forming the Guild was explicitly cultural: we collectively thought that hierarchies are not the best way to construct a mutually beneficial organization. If our Guild has any shape at all, it is a network, with true peer-to-peer cooperation.

Guild principals have authored more than thirty books. Their previous collaboration, *Adrenaline Junkies and Template Zombies* –

Understanding Patterns of Project Behavior, won the Jolt Award for best general computing book.

The Guild covers six time zones with principals living in the United States, Germany, and the United Kingdom. We are all fond of good food and good wine.

Steve McMenamin was a Principal of The Atlantic Systems Guild from its founding in 1983 until his death in 2019. He had a parallel career as a manager at Edison International, Crossgain, and later at Hawaii Electric. Over his lifetime he managed more than a thousand people and was known for taking a continuing and compulsive interest in the advancement of their careers.

Suzanne Robertson is a rock star in the systems engineering and socio-technical worlds. She has co-authored several best-selling books, including *Mastering the Requirements Process*, a guide to discovering and communicating the real problem. She has travelled the world teaching her courses and speaking at conferences. Her off-duty interests include opera, cooking, skiing and finding out about curious things.

Tim Lister is a Principal of the Atlantic Systems Guild based in New York City. He is a self-proclaimed Risk Management Zealot for our industry. He is convinced that the earliest stage of a project predicts all of the future for that project. Get it right, and you could have a ripping success. Muddle early; muddle for the duration.

Peter Hruschka dedicates his working life to technology transfer, especially for large embedded systems. He is the co-creator of arc42 and req42 – the homes for pragmatic software architecture and agile requirements engineering. Peter has published more than a dozen books and numerous articles on software and system engineering and teaches and consults worldwide.

James Robertson is a problem solver, consultant, teacher, photographer, author and practitioner of systems and software solutions. He is co-author of seven books and the Volere approach to requirements engineering. In a former life he was an architect, and so brings the architectural ideas of space, function, flow, and appearance to the design of business solutions. He has consulted and worked in every continent.

Tom DeMarco is the author or co-author of sixteen books including *Peopleware*, *Slack*, books on software development and management, five novels, and a collection of short stories.

Printed in Great Britain
by Amazon